AF545123

HAZARDOUS WASTE TREATMENT TECHNOLOGIES

HAZARDOUS WASTE TREATMENT TECHNOLOGIES

Biological Treatment, Wet Air Oxidation, Chemical Fixation, Chemical Oxidation

by

Alan P. Jackman
Robert L. Powell

Department of Chemical Engineering
University of California, Davis
Davis, California

NOYES PUBLICATIONS
Park Ridge, New Jersey, U.S.A.

Library of Congress Catalog Card Number: 90-24375
ISBN: 0-8155-1268-6
Printed in the United States

Published in the United States of America by
Noyes Publications
Mill Road, Park Ridge, New Jersey 07656

10 9 8 7 6 5 4 3 2 1

Library of Congress Cataloging-in-Publication Data

Jackman, Alan P.
Hazardous waste treatment technologies : biological treatment, wet air oxidation, chemical fixation, chemical oxidation / by Alan P. Jackman, Robert L. Powell.
p. cm.
Includes bibliographical references and index.
ISBN 0-8155-1268-6 :
1. Hazardous wastes--Purification. I. Powell, Robert Leslie, 1951- . II. Title.
TD1060.J33 1991
628.4'2--dc20 90-24375
CIP

Acknowledgments

This book results from work supported by the Alternative Technology Section of the Toxic Substance Control Division of the California Department of Health Services. We gratefully acknowledge the patience and the encouragement of Kim Wilhelm and Pam Johnson. We also thank Megan Taylor and S. Kent Stoddard of the California State Assembly who urged us to pursue these studies.

The work would not have been completed without the active participation of Drs. Chen Chen Fu and Pak Leung. They served as the last and most long lasting of a team of students and visiting scholars known as the Toxics Reduction and Analysis Project (TRAP). Many "TRAPers" are responsible for portions of this work. Two who stand out are Eleazar Ruimy, who organized the database and developed the data analysis strategy; and Jack Baylis who provided a steady hand over several years in many capacities.

We also thank the staff of the Department of Chemical Engineering for their support of this project, especially Lynn Bedillion and Darby Rivette for the long hours and tedious work involved.

January, 1991
Davis, California

Alan P. Jackman
Robert L. Powell

NOTICE

To the best of the Publisher's knowledge the information contained in this publication is accurate; however, the Publisher assumes no responsibility nor liability for errors or any consequences arising from the use of the information contained herein. Final determination of the suitability of any information, procedure, or product for use contemplated by any user, and the manner of that use, is the sole responsibility of the user.

The book is intended for informational purposes only. The reader is warned that caution must always be exercised when dealing with hazardous wastes. Expert advice should be obtained at all times when implementation of treatment technologies is being considered.

Mention of trade names, commercial products, or suppliers does not constitute endorsement or recommendation for use by the Publisher.

All information pertaining to law and regulations is provided for background only. The reader must contact the appropriate legal sources and regulatory authorities for up-to-date regulatory requirements, and their interpretation and implementation.

Contents and Subject Index

1. Introduction

This book seeks to address two outstanding issues in the reduction and regulation of hazardous wastes. Firstly, there are many facilities which currently produce hazardous wastes that are either sent to land disposal or sewered with little or no on-site treatment. Often such practices result from a lack of information concerning available technologies which might be applied to a particular spectrum of wastes, and cost estimates for implementing those technologies. In such instances, it is important that facility managers are aware of the best available and proven technologies which can be used to treat their wastes. It is also important that an independent source verifies that similar wastes have been treated using the same technology. In the cases of four hazardous waste treatment technologies -- biological treatment, wet air oxidation, chemical fixation and chemical oxidation -- we have compiled the information necessary to determine the applicability of a particular technology to a variety of wastes. The scientific basis for the treatment methodology is presented, along with representative process flowsheets which can be used under a variety of treatment-demand conditions. Cost estimate algorithms are presented along with list of suppliers of the technologies. We have attempted to provide an accurate assessment of the types of wastestreams which might be treated by examining a large database containing information on hazardous waste generation for the State of California based upon 1985 Biennial Generator Report Forms.

These Biennial Generator Reports also provide the means to fulfill the second goal of this book, which is to aid government planners in their decision-making process. The 1985 California survey has generated one of the largest, most detailed databases on hazardous waste generation. Such databases are becoming increasingly available [1], but pose special problems in their interpretation and use. In Chapter 2, we discuss the details of transforming the information collected into a self-consistent computerized database. Next, we consider the ancillary data needs associated with using such a database to evaluate treatment alternatives for particular wastestreams. For example, we discuss the means to evaluate whether or not a particular wastestream can be incinerated. We contrast the realities imposed upon the evaluation process by an actual data set against the idealized situation in which highly detailed physical and chemical data would be available. A scheme is then described which permits self-consistent *ad hoc* assumptions to be made which can provide realistic estimates of the impact of treatment alternatives on hazardous waste disposal patterns. This methodology is then systematically applied to the four treatment technologies reviewed in the subsequent chapters. The results of these studies provide realistic assessments of the potential impact of each technology in a large, industrialized state. They also give a wastestream-by-wastestream assessment of the applicability of a particular technology.

The 1976 Resource Conservation and Recovery Act (RCRA) coupled with the 1984 Hazardous and Solid Waste Amendments have imposed stringent and broad regulations on hazardous wastes, particularly requirements for pretreatment prior to land disposal, and wastes generated by small quantity generators (those producing 100 to 1000 kg/month) [2,4]. Robbins [2] has further stated that in the upcoming Congressional debates on RCRA reauthorization, reducing hazardous wastes will be of prime concern [4]. Under existing regulations, land disposal of wastes will clearly decline as fewer facilities are licensed. Highly toxic wastes (dioxin containing wastes, liquids with free cyanides, polychlorinated biphenyls, wastes containing halogenated organic compounds) have been banned since 1987. Now, however, all hazardous wastes listed by the Environmental Protection Agency are in the process of being banned from land disposal unless the generator can demonstrate that a practical treatment alternative is not available [2].

These regulations, as well as those applying to small quantity generators [3] represent an increased need for accurate information on alternative treatment technologies. Both industry representatives and regulators must be able to swiftly and accurately assess the efficacy of a treatment process for a particular hazardous wastestream. Such a "first cut" capability would allow even small-sized firms without adequate human resources to address issues being raised by the regulators. They, in turn, would be able to assist the generators by providing practical information as to the usefulness of a treatment method and its costs.

REFERENCES

1. "Hazardous Waste Management Database Starts to Take Shape" Chemical & Engineering News, pg. 19, February 9, 1989.
2. Robbins, A. Notes from presentation given on April 18, 1989, U.S. EPA Region 9.
3. Schwartz, S.I., and Cukovich, W.P., Managing Hazardous Wastes of Small Quantity Generators, Island Press, Washington, D.C.
4. Solving the Hazardous Waste Problem - EPA's RCRA Program, U.S. Environmental Protection Agency, Office of Solid Waste, Washington, D.C 20460, November 1989, EPA/530-SW-86-037, pg. 18-22.

2. Hazardous Waste Data and Their Analysis

2.1 DATA

2.1.1 Hazardous Waste Data

While our principal objective is to present information on available hazardous waste treatment technologies, we also wish to evaluate the potential impact of these technologies on typical hazardous wastes. To achieve this, we have used data available from the State of California which were collected as part of the 1985 Biennial Generator Report (BGR). These data were gathered by the Alternative Technology Section of the Toxic Substances Control Division of the Department of Health Services. The written data were transferred to the Department of Chemical Engineering at the University of California at Davis for validation and entry into a computer database. This chapter briefly describes these data, their format, and a typical technology analysis which can be performed using them. In the last case, we discuss the usefulness of incineration for disposing of hazardous wastes generated in California.

The data collected in the 1985 California Biennial Generator Report were unique in that they were tied to individual wastestreams within a facility. That is, every facility generating two tons of hazardous waste per year in the State was sent a Biennial Generator Report form to complete. The instructions with the form stated that one form was to be filed for each wastestream that surpassed the two ton per year limit. Some large facilities filed over fifty Reports. About 10% of the facilities responded to the questionnaire: 2470 facilities filed 9938 Reports.

Most of the information containe_ on these Reports is listed in Table 2.1. Some of the information is self-descriptive. The Environmental Protection Agency Identification Number (EPA ID) is the unique alphanumeric code associated with each registered hazardous waste generator, or treatment-storage-disposal (TSD) facility. The Wastestream Category is the same as the California Waste Code (CAL Code). This is a three-digit numeric identification used on hazardous waste manifests to identify the waste. Table 2.2 lists CAL Code definitions. The Wastestream Description is the narrative description associated with a particular CAL Code, for example, CAL

Table 2.1 Information Contained on 1985 Biennial Generator Reports

Company Name
Environmental Protection Agency Identification Number
Wastestream Category
California Waste Code (CAL Code)
Generating Process
Standard Industry Code (SIC)
Approximate Annual Volume
Major Components/Hazardous Constituents and Their Concentrations
Physical State (Liquid, Gas, Sludge, Solid)

Code 134 is an "Aqueous solution with total organic residues less than 10 percent." The "SIC Code" is the Standard Industry Code which identifies a facility according to the type of industrial activity being pursued. Table 2.3 lists SIC code definitions. In our study, only two digit SIC Codes were used. The "Approximate Annual Volume" roughly gives the generating rate of waste. It is, however, deceiving to interpret this too narrowly. Often, wastes may be due to a one time event at a facility, such as contaminated soil resulting from the clean-up of an unintended release of a hazardous material or waste, or, discarded, contaminated machinery associated with the retooling of an industrial facility. In our analyses, we have assumed that while disposal of such wastes may be a one time occurrence at a particular facility, such wastes will continue to be generated due to events and capital improvements at other facilities. The "Major Components/Constituents" are, ideally, the chemical names of the compounds present in the wastestream. Generally, however, it was found that such information was not available and that considerable follow-up effort was required. A detailed discussion of this activity is given by Bell, Jackman and Powell [1]. Similarly, the "Concentration" of each constituent was generally not given in precise terms.

A complete description of the procedures used for validating the data can be found in the report by Bell, Jackman and Powell [1]. Keyboard entry was used to input the data to a database, referred to hereafter as the BGR database, established on an International Business Machines Personal Computer/AT. The software used for the database was dBase III$^+$ (Ashton-Tate). Searches could be performed on any of the fields described in Table 2.1. Of particular interest were searches according to chemical constituent. These data were also the most difficult to obtain in a consistent format. To the extent possible, the classification suggested by Chemical Abstracts Service was used. However, there were many ill-defined "constituents" which, based upon conversations with the generating companies and our knowledge of the generating process, we defined somewhat arbitrarily. For the purposes of the analyses presented here, such definitions will lead to uncertainties in the absolute volumes which can be treated by a particular process. However, because of the large number of wastestreams represented by our sample (nearly 10,000), we believe that on a statistical basis, our results represent a good qualitative view of the spectrum of wastes in a large state with a diverse industrial base (California).

2.1.2 Property Database

To analyze the data from the Biennial Generator Reports, additional information on chemical properties was required. This first became apparent when we began to examine the prospective impact of high temperature incineration on hazardous waste disposal. Here, it is necessary to estimate the heat of combustion of a wastestream to determine the amount of make-up fuel required. It is also necessary to know the full spectrum of wastes in a particular stream so that any post processing of the flue gases can be considered.

A property database was created to allow treatment technologies to be analyzed. The elements of this database are listed in Table 2.4. It is necessary to note that not all of the elements were eventually completed, and also, that no existing database was found which was adequate for the project at hand.

The “Component Name” is the link between this property database and the BGR database. In some cases, it was found that component names were not unique and so a list of component name “Synonyms” was completed and included in the property database. For example, the

Table 2.2 Definitions of California Hazardous Waste Codes (CAL Codes)

Code	Definition
Inorganics	
111.	Acid solution $2 \le pH \le 7$ with metals (antimony, arsenic, barium, beryllium, cadmium, chromium, cobalt, copper, lead, mercury, molybdenum, nickel, selenium, silver, thallium, vanadium, and zinc)
112.	Acid solution without metals
113.	Unspecified acid solution
121.	Alkaline solution ($pH \ge 12.5$) with metals (see 111.)
122.	Alkaline solution without metals
123.	Unspecified alkaline solution
131.	Aqueous solution ($2 < pH < 12.5$) containing reactive anions (azide, bromate, chlorate, cyanide, fluoride, hypochlorite, nitrate, perchlorate, and sulfide anions)
132.	Aqueous solution with metals (see 111.)
133.	Aqueous solution with total organic residues 10 percent or more
134.	Aqueous solution with total organic residues less than 10 percent
135.	Unspecified aqueous solution
141.	Off-specification, aged, or surplus inorganics
151.	Asbestos-containing waste
161.	FCC waste
162.	Other spent catalyst
171.	Metal sludge (see 111.)
172.	Metal dust (see 111.)
181.	Other inorganic solid waste
Organics	
211.	Halogenated solvents (chloroform, methyl chloride, perchloroethylene, etc.)
212.	Oxygenated solvents (acetone, butanol, ethyl acetate, etc.)
213.	Hydrocarbon solvents (benzene, hexane, Stoddard, etc.)
214.	Unspecified solvent mixture
221.	Waste oil and mixed oil
222.	Oil/water separation sludge
223.	Unspecified oil-containing waste
231.	Pesticide rinse water
232.	Pesticides and other waste associated with pesticide production
241.	Tank bottom waste
251.	Still bottoms with halogenated organics
252.	Other still bottom waste
261.	Polychlorinated biphenyls and material containing PCBs
271.	Organic monomer waste (includes unreacted resins)
272.	Polymeric resin waste
281.	Adhesives
291.	Latex waste
311.	Pharmaceutical waste
321.	Sewage sludge
322.	Biological waste other than sewage sludge
331.	Off-specification, aged, or surplus organics
341.	Organics liquids (nonsolvents) with halogens
342.	Organic liquids with metals (see 111.)
343.	Unspecified organic liquid mixture
351.	Organic solids with halogens
352.	Other organic solids

Table 2.2 Definitions of California Hazardous Waste Codes (CAL Codes) continued

Sludges	
411.	Alum and gypsum sludge
421.	Lime sludge
431.	Phosphate sludge
441.	Sulfur sludge
451.	Degreasing sludge
461.	Paint sludge
471.	Paper sludge/pulp
481.	Tetraethyl lead sludge
491.	Unspecified sludge waste
Miscellaneous	
511.	Empty pesticide containers 30 gals. or more
512.	Other empty containers 30 gals. or more
513.	Empty containers less than 30 gals.
521.	Drilling
531.	Chemical toilet waste
541.	Photochemicals/photoprocessing waste
551.	Laboratory waste chemicals
561.	Detergent and soap
571.	Fly ash, bottom ash, and retort ash
581.	Gas scrubber waste
591.	Baghouse waste
611.	Contaminated soil
612.	Household wastes

Table 2.3 Summary of SIC Codes

Code	Description
Agriculture	
07	Agricultural Services
Mining	
10	Metal Mining
13	Oil and Gas Exploration
Construction	
20	Food and Kindred Products
24	Lumber and Wood Products
25	Furniture and Fixtures
26	Paper and Allied Products
27	Printing and Publishing
	2711 Newspapers
28	Chemicals and Allied Products
	2810 Chemicals, industrial inorganic
	2820 Plastic materials & synthetics
	2830 Drugs
	2840 Soaps, cleaners & toilet goods
	2851 Paint & allied products
	2860 Chemicals, industrial inorganic
	2870 Chemicals, agricultural
	2890 Chemical products, miscellaneous
29	Petroleum Refining and Related Industries
	2911 Petroleum refining
	2950 Paving & roofing
30	Rubber and Misc. Plastic Products
31	Leather Tanning and Finishing
32	Stone, Clay and Glass Products
	3210 Glass & glassware
	3270 Concrete, gypsum & plaster products
33	Primary Metal Industries
	3320 Foundries, iron & steel
	3330 Nonferrous metals
34	Fabricated Metal Products
	3410 Cans & shipping containers
	3430 Plumbing & heating
	3460 Forging & stamping
	3470 Metal services
35	Machinery, except electrical
36	Electrical and Electronic Machinery
	3674 Semiconductor & related devices
37	Transportation
	3710 Motor vehicle
	3720 Aircraft & parts
	3730 Ships & boats
	3760 Guided missiles & space vehicles
38	Instruments and Related Measuring Dev.
39	Misc. Manufacturing Industries
Transportation	
40	Rail Transportation
42	Trucking
49	Electric, Gas, and Sanitary Services
Wholesale/Retail Sales	
73	Business Services
75	Automotive Repairs
95	Governmental Agency, Except Military
97	Governmental Agency, Military
99	Not Otherwise Specified

component "acrylonitrile" has as synonyms "cyanoethylene," "fumigrain," "ventox," and "vinyl cyanide."

Table 2.4 Elements Included in Property Database.

Alphanumeric Fields	*Logical Fields*
Component Name	Acid/Base
Synonym(s)	Organic
Chemical Formula	Highly Toxic
Molecular Weight	Azeotrope
Boiling Point	
Heat of Vaporization	
Heat of Combustion	
Incineration Restrictions	
Other Restrictions	
Chemical Abstracts Service Number	
Price	

The "Chemical Formula" and "Molecular Weight" are frequently needed to determine appropriate alternatives. For example, the presence of fluorine, chlorine, bromine or iodine places restrictions on options for treatment and disposal of wastes. To evaluate the usefulness of incineration, we included several specific fields in the database including the "Heats of Combustion" and "Vaporization," the "Boiling Point" and a series of codes which would signal that special care would be required for the effective and safe incineration of a wastestream. These characteristics are summarized in Table 2.5. "Other Restrictions" refer to characteristics of a component which signal that it becomes dangerous when one or more of the following environments are present: (1) heat, (2) acid or base, or (3) water. The "Chemical Abstract Registry Number" was included so that, if necessary, more information about the component could be obtained from commercial databases. The "Price" field was included so that the treatment analyses could include an economic component.

Table 2.5 Special Chemical Characteristics

Characteristic	Code	Characteristic	Code
Aromatic	ARO	Mercaptan	MER
Asbestos	ASB	Nitro organics	NHR
Cyanide	CNR	Organic carbon compounds	ORG
Dibenzofuran	DIB	Polychlorinated biphenyls	PCB
Halogenated compounds	HAL	Phenol	PHE
Heavy metal compounds	HME	Quinone	QUI

The logical fields included indications as to whether the component is an "Acid" or "Base;" and whether it is an "Organic Compound." The "Highly Toxic" field identifies compounds with toxicity so high that special handling may be necessary. The components included in this category result from the definitions of toxic waste established by the Environmental Protection Agency in 1980 and can be found in Brunner [2]. The last logical field listed in Table 2.3 distinguishes the components that form azeotropes with water from those that do not. This information can be important in judging whether components of the stream can be separated by distillation. Many streams contain a high percentage of water, and those components which form azeotropes can pose special problems if distillation is used as part of the treatment process. This information as well as the other information required to complete this database can be found in standard references [3-6].

One of the largest problems in this or similar studies results from the collection of chemical component data from generators, who may not employ personnel trained in chemistry or may not have access to specific chemical information about their wastestreams. Such "component" information may be vague at best, and, at worst, misleading. To handle such cases, we have defined a class of "components" which we call generic. Such substances are usually mixtures, and, not infrequently, the mixtures are complex and ill-defined. For example, in the case of oils, the Biennial Generator Survey yielded over 200 apparently distinct classifications. Many of the names described oils with very similar properties, although there is generally a wide range of properties associated with oils. Rather than attempting to assign different properties to each oil listed, the 200 different oils were divided into three categories which span the range of properties. The viscosity of the oil was used as the property on which to base the differentiation among the oils. As a purely practical consideration, this property was chosen because it was feasible, and from a more technological point of view, because of its impact upon the design of treatment facilities, particularly incinerators. All of the oils were given one of three (arbitrary) component names: Oil 1, Oil 2 or Oil 3. The chemical properties of each oil were determined by making assumptions as to the typical compositions of the distinct components.

A total of 69 generic substances were defined. These substances were able to represent, at least approximately, the properties of several hundred components actually found on the Report Forms. A list of these substances is given in Table 2.6. These substances range from chemical names such as "alcohol" to such catchall categories as "construction waste 1." In all cases, generic compounds have been defined while trying to maintain the most relevant features of the information contained in the original component name. For example, the possible types of wastes in the category "Containers and Equipment" are endless. To minimize this problem, this category was subdivided into six subcategories, designated by the alphanumeric codes A1, A2, B1, B2, C1, C2. The number indicated whether the container or equipment was combustible (1) or not combustible (2). The relative size of the wastestream was categorized by a letter: "A" indicated a large container or piece of equipment (volume of 55 gallons or more); "C" indicated a volume less than 5 gallons; "B" indicated those volumes which were in-between.

Upon Examining Table 2.6, it can be observed that generic substances occur in two forms:

1. They appeared on the Biennial Report Forms as a "chemical component" but they clearly represented mixtures. "Abrasive Material" and "Battery (Automotive)" are examples of this.

Table 2.6 List and Description of Generic Substances

Basic Component Names	Generic Substance Description
ABRASIVE MATERIAL	All grinding compounds
ABSORBENT 1	Organic absorbents (sawdust)
ABSORBENT 2	Inorganic absorbents and those not specified
ACID SOLUTION 1	Strong acids (pH 0-2)
ACID SOLUTION 2	pH range of 2-4 plus those not specified
ACID SOLUTION 3	pH range of 4-7
ADHESIVE (LATEX)	Water-based adhesives
ADHESIVE (EPOXY)	Epoxy-based adhesives
ADHESIVE (ACRYLIC)	Acrylic-based adhesives
ALCOHOL	Generic alcohols
ALIPHATIC HYDROCARBON	Aliphatics, carbon chain, alkanes hydrocarbon, olefin
ALKALINE SOLUTION 1	pH range 7-9
ALKALINE SOLUTION 2	pH range 9-12 plus unspecified
ALKALINE SOLUTION 3	pH range 12-14
ALUMINUM	Aluminum chips, ducts, and traces
AROMATIC HYDROCARBONS	Alkyl benzene, aromatics
ASBESTOS	All asbestos containing wastes
ASH (FLY)	All ash waste
ASPHALT	All asphalt related waste
BIOLOGICAL WASTE	All bi-organic waste
CARBON	C, coal, coke, soot, toner
CATALYST	All general catalysts
CHLORINATED HYDROCARBON CLOTH	Clothing, rags, towels, etc.
CONSTRUCTION WASTE 1	Organic, combustible (wood)
CONSTRUCTION WASTE 2	Inorganic, noncombustible (brick, dirt, concrete)
CONTAINER AND EQUIPMENT (1/2)&(A/B/C)	"1" - combustible "2" - noncombustible "A" - very large "B"- >5 and <55 gals or medium "C" - small
DETERGENT AND SOAP	All detergents and soaps
ETHYLENE GLYCOL	Coolants and antifreezes
EXPLOSIVE	Unknown explosives
FATTY ACID GLYCERIDE	Fats, fatty acids
FLUX	Flux (alcohol, residue, rosin, solder)
FUEL 1	Aviation fuel, gas, kerosene
FUEL 2	Diesel, residue
GREASE	Grease, oil (grease)
INK 1	Organic ink
INK 2	Inorganic ink
INK 3	Water-based, sludges
INK 4	Solid, residue
METAL MIXTURE 1	Inert metal mixtures (Fe, Ag, Al)
METAL MIXTURE 2	Heavy metals & toxic mixtures (Pb,Cr,Ni,Hg,Th,Zn)
OIL 1	Light oils
OIL 2	Medium weight oils (cutting)
OIL 3	Heavy oils (crude, tars)
ORGANIC ACID	All unspecified organic acids
PAINT 1	Organic, combustible paints and lacquers
PAINT 2	Paints with metal pigments

Table 2.6 List and Description of Generic Substances (continued)

PAPER	All paper products
PESTICIDE ( )	Pesticides (name)
PCB CONTAMINATED SOLIDS	Equipment using PCB's
PHOTOCHEMICALS	All photochemical waste
PIGMENT 1	Organic pigments
PIGMENT 2	Inorganic pigments
PIGMENT 3	Pigments containing heavy metals
PLASTIC	Plastics, polyethylene, resins, polyurethane, Polymers
RUBBER	Gloves, foam, liners, synthetic
SEWAGE	Sewage
SLUDGE	Unspecified sludge
SOIL AND SAND	Soil and sand
SOLVENTS	Solvent (blend, bulk, combustible, flammable), thinner
TRASH 1	Highly combustible (cork, cardboard, organic matter)
TRASH 2	Medium combustible (debris, gloves, equipment)
TRASH 3	Non-combustible trash
WAX (LIQUID)	Floor waxes
WAXES	General waxes

2. Components for which properties might have been found were lumped with non-specific components to make a generic substance. An example of this is "Photochemicals." The specific components, ammonium silver thiosulfate and sodium silver thiosulfate, were categorized under this category, and were ignored as individual components.

If the generic substance included chemical compounds having a very wide range of properties, then several generic names were defined, each representing mixtures of chemical compounds having a smaller range of physical properties, as discussed in the examples above. Finding or determining physical and chemical properties for generic substances was easier for some than others. For example, it was relatively easy to define the "Oil" categories, while it was difficult to define "Absorbent 1" and "Pigment 1." Most absorbent reported was inorganic material, which was called Absorbent 2. Absorbent 1 includes absorbents which themselves are organic. Properties of organic absorbents could not be found in any available sources. Similarly, in the case of pigments, "Pigment 1" is the category for organic pigments. However, the majority of the pigments described in the literature are composed of inorganic materials. The definitions which were chosen for the absorbents and pigments were convenient for assessing the incinerability of wastestreams, although little specific information about the properties could be found.

Since generic substances were mixtures, there could be no precise molecular structure assigned. The chemical formula was assigned in one of the following ways:

1. The formula was determined as the weighted average of the formulas of the representative compounds for that generic substance. An example is "Pigment 2," which has titanium oxide, zinc oxide and calcium carbonate in its formula.
2. The formula was determined as that of a single compound whose properties were thought to be highly representative of those of the generic substance. For "Aromatic Hydrocarbons," toluene and benzene were chosen as the representative components. In the case of "Chlorinated Hydrocarbon," the formula for chloroform was used.

The molecular weight of a generic substance was related to the molecular weights of the representative compounds. These were determined by one of the following methods:

1. The formula chosen for the generic substance itself when it was represented by a chemical compound.
2. An average value of the molecular weights of the highest percent components. This was applicable to those generic substances containing higher concentrations of some representative components relative to others. For example, since "Paint 1" contained large concentrations of 1,1,1-trichloroethane and trichloroethylene, its molecular weight was fixed using a weighted average of the molecular weights of these components.
3. An average value of the molecular weights of the representative compounds. This was applied to those generic substances with constituents containing relatively similar molecular weights. For example, with "Pigment 2" the molecular weights of titanium oxide, zinc oxide and calcium carbonate were averaged.

The classification of any compound in the generic substance as "highly toxic," "strong acid," "strong base," or as a substance with processing restrictions, or as a substance having an azeotrope with water, or as an ignitable substance determined the classification of the generic substance. Values of the boiling point, heat of vaporization and heat of combustion were assigned such that conservative decisions regarding incinerability would result. The highest boiling compound representing the generic category was chosen as the boiling point for those cases where large variations (greater than 20 K) in values occurred among the representative compounds. If the boiling points were close (within approximately 10 K), then an average value was assigned. The heat of vaporization was chosen as either the largest value or the average value of the heats of vaporization of the representative compounds. For those generic substances which were combustible, a heat of combustion was determined from either the heats of combustion of the representative compounds or using Brunner [2]. Two methods were used:

1. An average value was calculated using all of the representative compounds.
2. The lowest value was chosen depending upon the variation of the values (e.g., the variation was within 2000 kcal/kg). For example, for "Paint 1," the smaller heat of combustion for 1,1,1-trichloroethane was chosen over that of trichloroethylene.

These techniques for assigning properties to the generic components result, in many cases, from the "best educated guesses." In our view, further refinements in the quality of the property database for such substances would be premature in that the quality of the typical survey information which would be analyzed is quite poor.

2.2 INCINERATION ANALYSIS

2.2.1 Background

To demonstrate the usefulness of the hazardous waste information collected on the Biennial Generator Reports, we conducted a study to ascertain the potential impact of incineration technology on the disposal patterns of hazardous waste in California. This study consists of a global analysis of incineration on a wastestream-by-wastestream basis. We first formulated a flowsheet which permits us to analyze a wastestream for incinerability based upon either its characteristics or constraints imposed by regulatory agencies. Our goal was to calculate the operating parameters required to incinerate a wastestream, such as make-up fuel, as well as to analyze the need for the post-treatment of residuals resulting from the incineration process. This study can be viewed two ways. Firstly, it creates a framework for the global analysis of wastestreams. Since incineration is a well-defined technology, performing this analysis allows us to assess the limitations of the analysis process, including the BGR database and the Property database. This framework can be expanded to include other technologies, or, it can be used to analyze another group of wastestreams. Secondly, the results of this study can be considered as giving general guidelines for establishing the potential impact of incineration on a particular spectrum of wastestreams. If we assume that the wastestreams from California would be representative of those of most major industrial states (or countries), it would be possible to grossly estimate the percent of incinerable wastes and the potential environmental and economic impact of implementing incineration as a treatment technology. In carrying out this study, we have used some regulatory guidelines currently either implemented or being considered in California as discussed in Chapter 1. For example, recently proposed legislation [7] would require any wastestream with a heating value greater than 1700 kcal/kg (3000 BTU/lb) to be incinerated or otherwise treated prior to land disposal. Another restriction soon to go into effect requires that any wastestream containing more than 1% volatile organic compounds must be treated similarly.

These two restrictions led us to generally set 1700 kcal/kg as the minimum heating value for the incineration of a wastestream. The exceptions were those streams containing more than 1% volatile organic compounds but having a heating value less than 1700 kcal/kg. Such streams were considered incinerable with supplementary fuel, although it is recognized that such a treatment methodology may not prove to be be the most effective economically. Even wastestreams having a heating value above 1700 kcal/kg are usually unable to sustain combustion without the addition of supplementary, high heat value fuels. The heating value needed for self-sustaining incineration was set at 5000 kcal/kg which resulted in the wastestreams being divided into three categories: can undergo self-sustained incineration, cannot undergo self-sustained incineration, cannot (or need not) be incinerated. Finally, a working definition of volatile organic compound was established as any component (with the exception of water) that has a boiling point less than 100°C.

2.2.2. Analysis

The flow sheet for the incineration analysis (program) is shown in Figure 2.1. This flow sheet actualizes the previous discussion in terms of the restrictions placed upon the applicability of the incineration process. Each wastestream, as defined by the Biennial Generator Report, is considered. The initial determination is the physical state of the wastestream. This determines the type of handling equipment which might be needed for the waste. After sorting according to the physical state, it is determined whether the stream is or must be incinerated (volatile organic compound present at a concentration greater than 1% or a heating value greater than 1700 kcal/kg) and whether that incineration will or will not be self-sustaining. The limits of 1% and 1700 kcal/kg could be readily changed to adapt to a different or new regulatory environment.

Typical results of this analysis are shown in Figure 2.2. All of the wastestreams which were reported on the Biennial Generator Report Form as being disposed off-site were analyzed. Most of the waste, 67%, was judged as being non-incinerable. This figure is even larger when the category "1% VOCs" (volatile organic compounds) is added. All such compounds in this category would fall under the heading of non-incinerable were it not for the hypothetical requirement discussed above. Less than 10% of the total wastes can undergo self-sustained incineration (heating value greater than 5000 kcal/kg) and somewhat more than 20% can be incinerated with supplementary fuel (heating value between 1700 kcal/kg and 5000 kcal/kg).

As seen in Table 2.1, the information contained in the Biennial Generator Reports allows other types of analyses to be performed. One of the most useful is to examine the waste profiles of particular industry groups and then ascertain the effect of treatment technologies on the wastes from that group. It may be possible, for example, to identify trends within an industry, such as decreases in waste generation or a change in the pattern of waste generation due to the

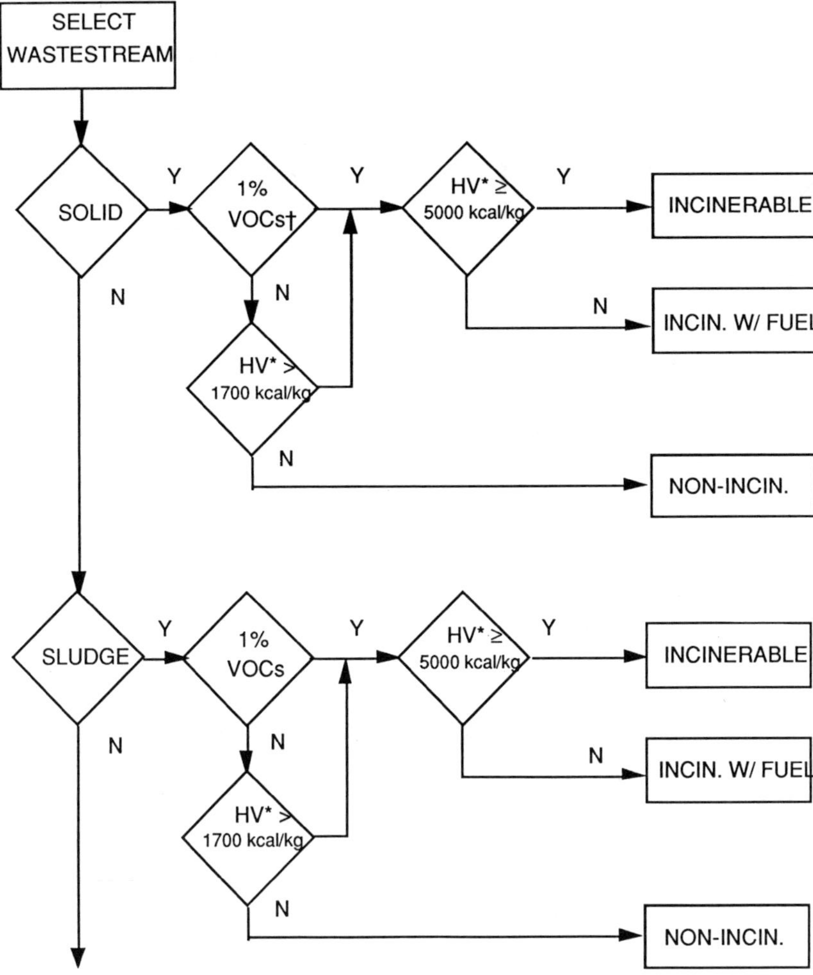

Figure 2.1. Flow Sheet for the Analysis of Incinerability of Wastes in the BGR Database (continued on following page).

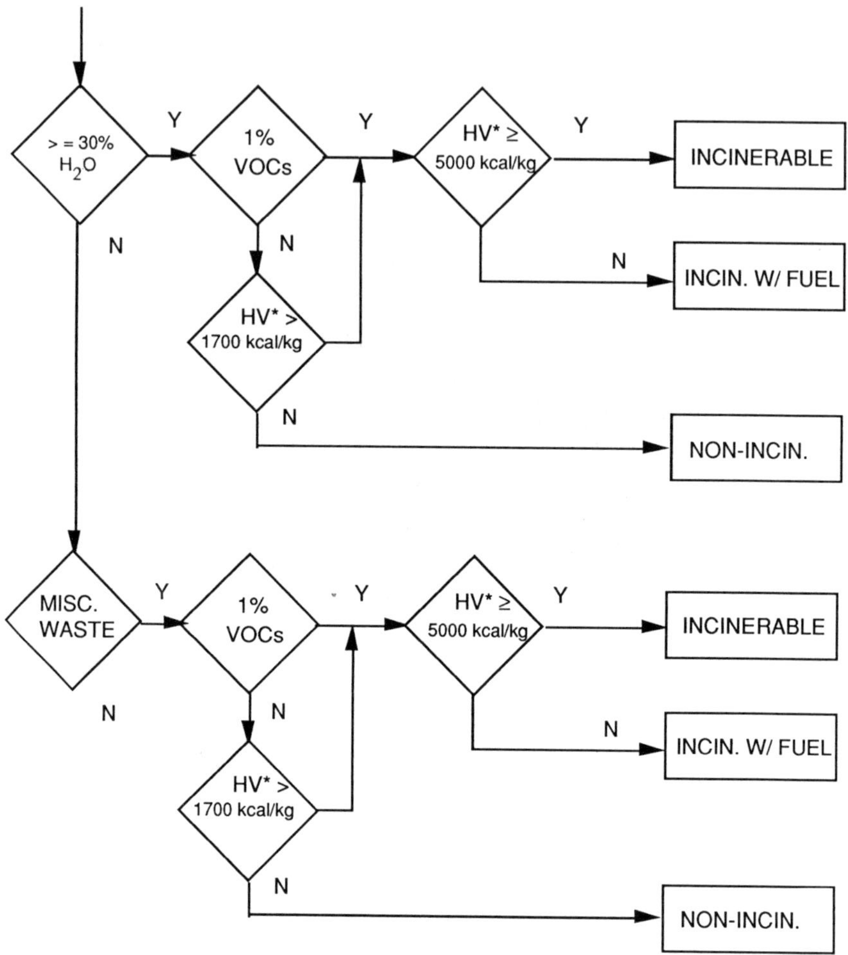

*Heating Value

† Volatile Organic Compounds

Figure 2.1. (continued)

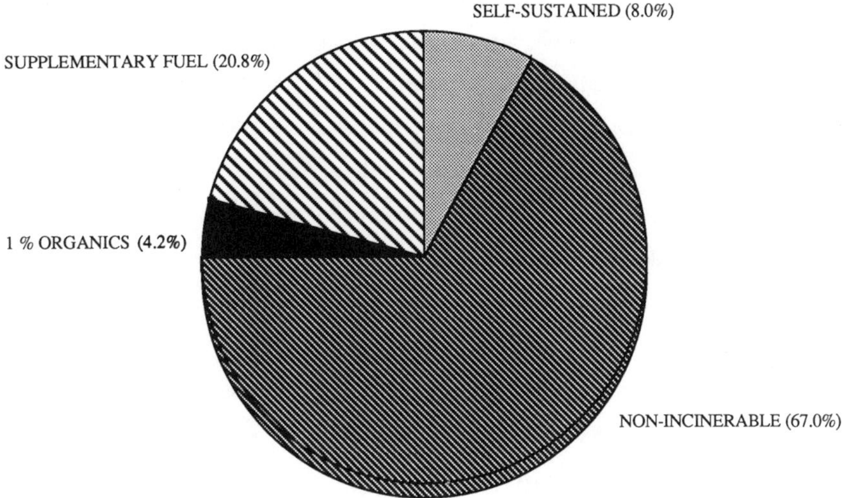

Figure 2.2. Incinerability of Waste in BRG Database by Categories

implementation of new technologies. The report of Bell, Jackman and Powell [1] details the results of such analyses for all industry groups as identified by two digit SIC Codes. Table 2.7 shows the results of such an industry group analysis for waste treatment by incineration. Here, it is found that SIC Code 29 (petroleum and coal products) accounts for most of the wastes which can undergo self-sustained incineration. The next highest producers of self-sustainable incinerable wastes are from SIC Codes 34 (fabricated metal products) and 40 (railroad transportation). The industries in both of these groups produce considerable quantities of combustible wastes such as waste oil and paper. The industries producing petroleum and coal products (SIC Code 29) also have the greatest volumes of wastes which can be combusted using supplementary fuels as well as the greatest quantities of wastes containing 1% volatile organic compounds.

The incineration analysis outlined in Figure 2.1 determines whether incineration of a waste stream is possible based upon its volatility and heat of combustion. After that judgement has been made it is necessary to determine if additional processing is needed after incineration. Such an analysis consists of: (1) determining the type of incineration for a waste stream, (2) deciding how to treat the flue gases after incineration, and (3) determining the types of treatment necessary for a stream which the analysis indicated was unsuitable for incineration.

The determination of the need for additional processing, if any, is focused on the constitutive make-up of the wastestream. Each component in the stream should be separately examined to determine its structure and its relation to the remainder of the stream. The stream can then be labeled according to the type of incineration and additional processing recommended. Certain components of a wastestream require specific decisions. For example, a wastestream containing polychlorinated biphenyls would require high temperature incineration. The atomic/molecular composition could reveal the presence of halogens which would require that scrubbers be used to treat the flue gas. Figure 2.3 is a flow sheet outlining the pathways that an appropriate analysis might follow. The information from such an analysis would be general, as the logic has been developed to look at large numbers of waste streams at one time and there is no provision for case-by-case studies. If the analysis suggests that additional processing is necessary for a specific wastestream, it should be followed with further investigations on a stream-by-stream basis. The overall purpose of such a study would be to provide a global picture of the types of incineration technology that are likely to be needed to treat a wide spectrum of hazardous wastes.

Table 2.7. Industries Generating the Largest Volume of Incinerable Waste

Rank	SIC	Volume Self-Sus Waste	Percent Self-Sus Waste
1	29	18,924	18.5
2	34	14,021	13.7
3	40	11,323	11.0
4	75	9,348	9.1
5	28	9,233	9.0
6	99	6,404	6.3
7	49	4,719	4.6

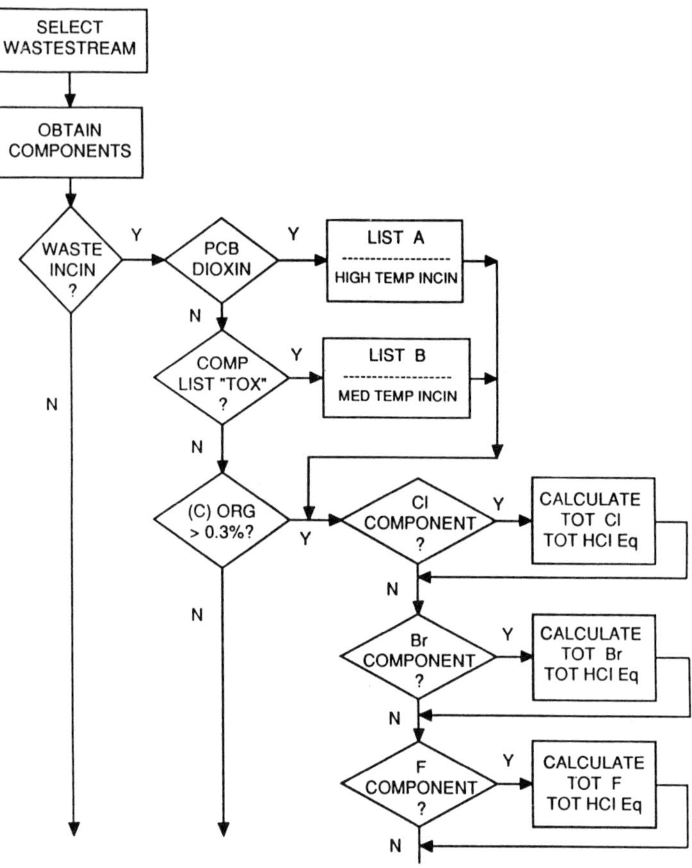

Figure 2.3. Flow Sheet for Analyzing the Need for Peripheral Treatment Associated with Incineration of Wastes in the BGR Database.

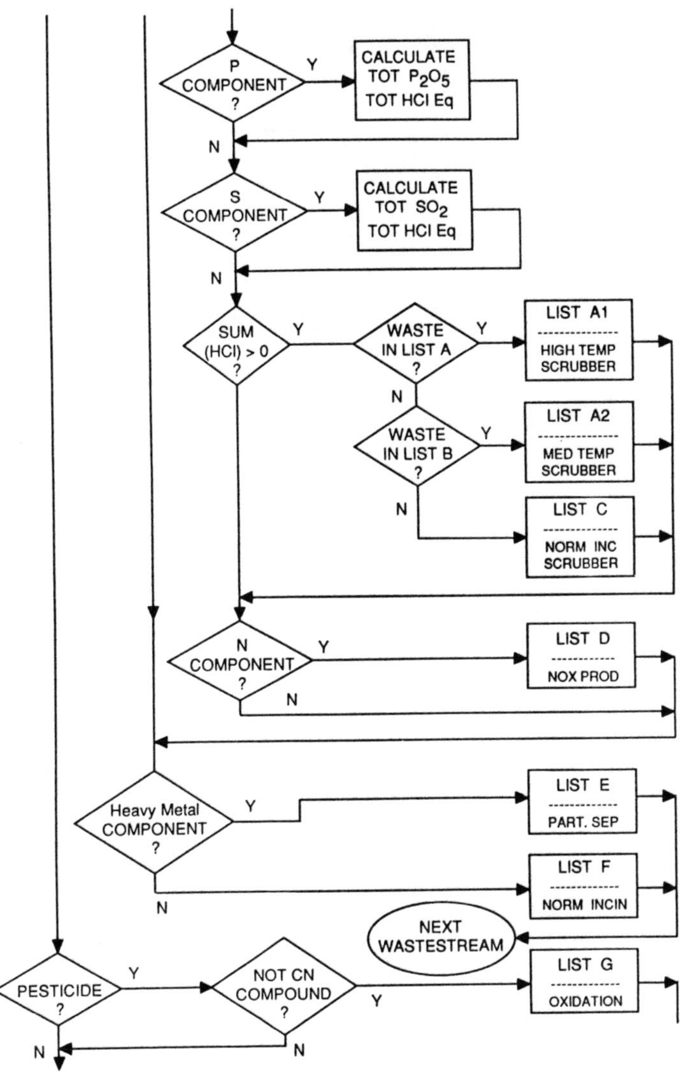

Figure 2.3 (Continued) Flow Sheet for Analyzing the Need for Peripheral Treatment Associated with Incineration of Wastes in the BGR Database.

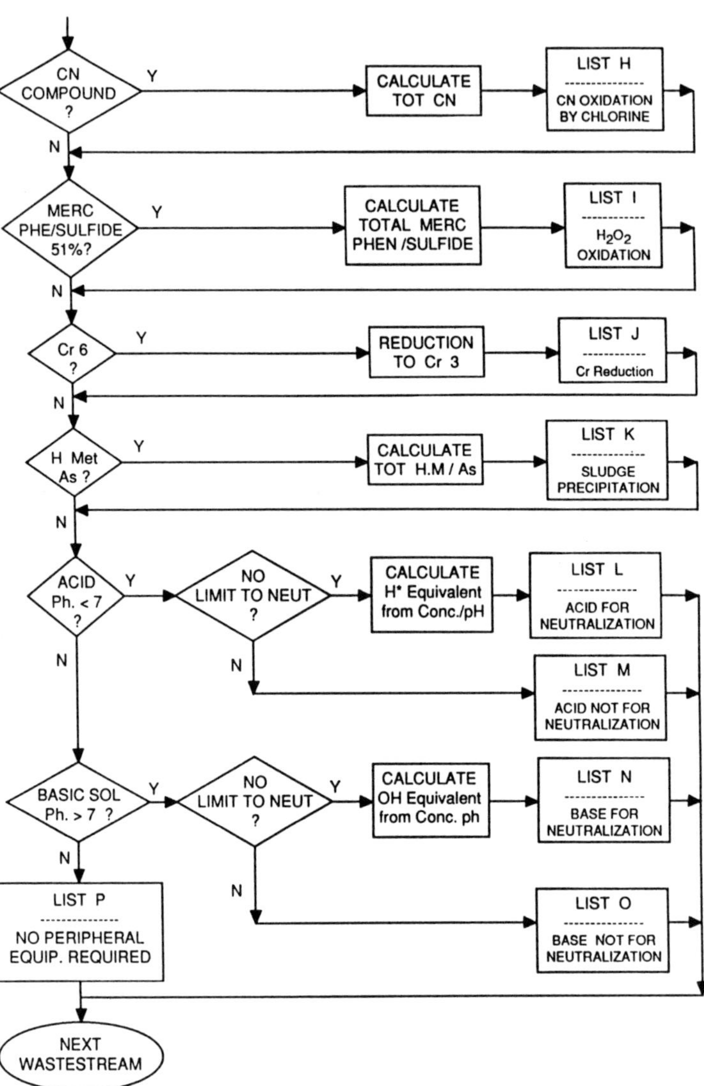

Figure 2.3 (Continued) Flow Sheet for Analyzing the Need for Peripheral Treatment Associated with Incineration of Wastes in the BGR Database.

2.3. REFERENCES

1. R. L. Bell, A. P. Jackman and R. L. Powell, a report submitted to the California Dept. of Health Services by Dept. of Chemical Engr., Univ. of California, Davis (1988).
2. Brunner, C.R., Incineration Systems, Selection and Design, Van Nostrand Reinhold Co. 1984.
3. Weast, R.C., Handbook of Chemistry and Physics, 51st Ed., Chemical Rubber Co., Ohio.
4. Lange, N.A. Ph.D., Forker, G.M., Handbook of Chemistry, 7th Ed., Handbook Publishers, Inc., Ohio, 1949.
5. Windholz, M., The Merck Index, 10th Ed., Merck & Co., Inc., N.J., 1983.
6. Chemical Hazard Response Information System (CHRIS), Department of Transportation and the Coast Guard, 1978.
7. Brunner, C.R., Incineration Systems, Selection and Design, Van Nostrand Reinhold Co. 1984.

3. Biological Treatment

3.1. INTRODUCTION

The biological treatment of wastestreams is based on the ability of a mixed population of microorganisms to utilize organic contaminants as nutrients. Organic constituents, which may be present either in solution, as a separate liquid phase, or as a solid, are removed by aerobically converting them to carbon dioxide and water (mineralization), or anaerobically decomposing them to methane and carbon dioxide, or biotransforming them to less toxic or non-toxic organic compounds.

Biological treatment processes do not alter or destroy most inorganics. Nitrogen and phosphorus are essential to cell growth and are consumed to some extent in biological waste treatment. Trace concentrations of inorganics may be partially removed from the liquid wastestream during the biological treatment, because of adsorption onto the microbial cell coating. Typically, microorganisms have a net negative charge and therefore are able to adsorb metal ions in solution by a cation exchange mechanism. Anionic species such as chlorides and sulfates are not affected by biological treatment. Therefore, wastestreams having organic constituents are the most likely candidates for biological treatment. This technique is widely used for treatment of municipal and industrial waste waters [1,2,3]. The application of biological treatment for in situ treatment of contaminated ground water is currently an active area of research and development.

The two major classes of biological treatment are aerobic (with oxygen) and anaerobic (without oxygen). The conventional application of biological treatment to industrial organic wastes employs techniques used for municipal waste treatment, e.g., activated sludge, trickling filters, rotating biological contactors, aerated lagoons, facultative lagoons, and anaerobic treatment. Several innovative applications of biodegradation techniques for treatment of hazardous waste have been developed, such as activated carbon adsorption with biological regeneration (PACT process [4]), fluidized-bed bioreactor [5], and the ICI deep shaft aeration system [6].

The BOD (Biological Oxygen Demand) and COD (Chemical Oxygen Demand) values are the most commonly used yardsticks to measure the pollutant load of a wastestream. The difference between the influent and effluent values of BOD or COD for a biological treatment process is used to determine the operating efficiency of the process. The BOD value is the amount of oxygen used by microorganisms to remove the organics in a wastestream. Because complete breakdown of even an easily biodegradable sample could take several weeks, a shorter standard incubation period (5 days at 20°C) is used to determine BOD value. However, this five day period is still too long to wait for making process control decisions that depend on pollutant loadings. The COD test, on the other hand, utilizes chemicals (e.g., potassium dichromate) rather than microorganisms to oxidize pollutants in the wastestream and requires only three hours to complete. Table 3.1 presents BOD and COD values of some toxic petrochemicals, and the stoichiometric oxygen demand for these chemicals, i.e., the theoretical amount of oxygen required for a complete oxidation to water and carbon dioxide[7].

Table 3.1 BOD and COD Values of Some Selected Compounds

Substance	S	BOD		COD	
	g/g	g/g	% of S	g/g	% of S
Allyl alcohol	2.21	1.79	81	2.12	96
n-Butyl alcohol	2.59	1.71	66	2.49	95
Diethylene glycol	1.51	0.05	3	1.51	100
Benzene	3.08	2.18	71	2.15	70
Styrene	3.08	1.29	42	2.80	91
Toluene	3.13	2.15	69	2.52	81
o-Xylene	3.17	1.64	52	2.91	92
m-Xylene	3.17	2.53	80	2.62	83
p-Xylene	3.17	1.40	44	2.56	81
Acetone	2.21	1.85	84	1.92	87
Acrolein	2.00	0.00	0	1.72	86
Phenol	2.38	1.68	70	2.33	98
3,5-Xylenol	2.60	0.10	4	2.60	100

S = stoichiometric oxygen demand
Source: [7]

The microorganism population in the biological treatment process can either be natural or developed to act on specific compounds in the waste. Both procaryotic and eucaryotic organisms have potential for biological treatment of toxic organics. Eucaryotes, which include protozoa, fungi and most groups of algae, have a highly organized cell structure. On the other hand, procaryotes, which include bacteria and blue-green algae, have a much simpler cell structure without a classical nucleus. A list of these microorganisms and their characteristics is presented in Table 3.2. Among those, bacteria are the most important to biological treatment. The other microorganisms, however, feed on dispersed bacteria and fine particulates that do not settle well and therefore help improve the quality of the effluent. Their role is secondary in importance to that of the bacteria.

The microbial population in a batch biological system passes through four distinct growth phases [8]. These phases are graphically illustrated in Fig. 3.1. After the waste and the microorganisms are combined, the reproduction rate of the microorganisms is quite slow, and the number of cells in the system increases very gradually. During this first growth phase, the microorganisms are becoming acclimated to their environment and starting to produce the enzymes they need to break down the organics present in the wastestreams. Following the initial acclimation period, the microorganisms enter a logarithmic growth phase. During this second growth phase, there is ample food (organic substrate) available, and the growth rate of the microorganisms increases significantly. As the substrate becomes limited, the growth rate declines and some microorganisms die. The dead microorganisms then become substrate for the remaining population. As the substrate continually decreases, the microorganisms metabolize the polysaccharide slime layer which coats their cell walls, resulting in denser and heavier cells. There is an optimum point at which the slime layer still exists and the organisms are fairly dense. At this point, if the organisms collide, they agglomerate and settle. However, if the substrate is so minimal that the slime layer has decreased to the point where there is no adhesion, the endogenous growth phase, the individual particles do not agglomerate. Floc formation in activated sludge treatment, and adhesion to support media in trickling filters and rotating biological contactors

Table 3.2 Names and Characteristics of Microbial Groups in Biological Treatment

Microorganism	Characteristics	Significance
Fungi		
Yeast	pH<5, ae/mae	Attacks and partially degrades complex compounds not readily metabolized by other organisms. Wide range of nonspecific enzymes.
Mold	pH<5, moisture about 50%	
Algae	ae/mae; light:600-700 nm; low carbon flux	Self-sustaining population, light is primary energy source, partially degrades certain complex compounds, photochemical reactions, oxygenates effluent, supports growth of other microbes, no aeration needed; effective in bioaccumulation of hydrophobic substances.
Cyanobacteria (formerly called blue-green algae)	ae/mae, an; light: 600-700 nm; low carbon flux	see algae
Bacteria		
Heterotrophs (aerobic)	ae; proper organic substrate	For many compounds degradation is more complete and faster than under anaerobic conditions. High sludge production.
Anaerobic (fastidious)	an	Conditions for abiotic or biological reductive dechlorination, certain detoxification reactions not possible under aerobic conditions; no aeration, little sludge produced.
Faculative anaerobes	mae/an	No aeration necessary, reductive dechlorination possible.
Photosynthetic bacteria		
Purple sulfur	an(light), mae(dark); light: 800-890 nm; low carbon flux	Self-sustaining population able to use light energy conditions right for reductive dechlorination, no aeration.
Purple nonsulfur	an; light: 800-890 nm; low carbon flux	See purple sulfur bacteria, also nonspecific enzymes.
Actinomycetes	ae, moisture: 80-87%, temp.: 23-28°C, urea as nitrogen source	Universal scavengers with range of complex organic substrates often not used by other microbes.
Oligotrophs (may be from almost any group above)	ae; carbon flux of <1 mg/L/d; favorable attachment sites	Removal of organic contaminants in trace concentrations, many inducible enzymes for multiple substrates.

ae = aerobic; mae = microaerophilic (<0.2 atm oxygen); an = anaerobic
Source: [3]

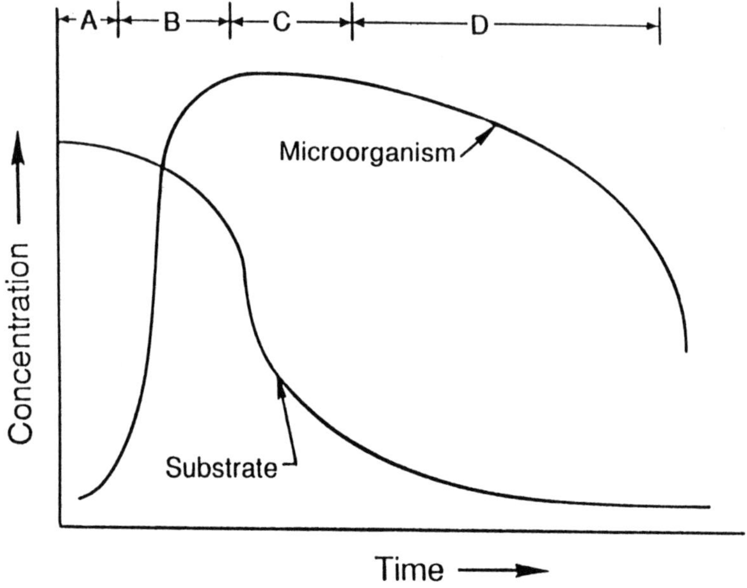

Figure 3.1. Biological Growth and Substrate Consumption.

rely on the biomass retaining their slimy coating, and are therefore best operated in the log-growth phase. Aerated lagoons and anaerobic digestion have long retention time and exhibit endogenous growth.

Because biological systems contain living organisms, they require specific ratios of carbon and nutrients. The most important nutrients are nitrogen, phosphorus, sulfur, potassium, calcium, and magnesium (macro-nutrients) [9]. Some nutrients are just as essential as those listed above, but they are needed only in trace amounts. These are iron, boron, copper, manganese, zinc, chromium, and cobalt (micro-nutrients) [9]. Water is also a necessary component of all living organisms and therefore is a vital part of the biological waste treatment systems. Industrial waste waters often lack the essential macro-and micro-nutrients which must therefore be added during treatment.

The body temperature of microorganisms is at or near the temperature of the medium in which they exist (poikilotherms). Microorganisms have a range of temperatures in which they achieve optimum growth. Temperature ranges for psychrophilic organisms are 0°C to 20°C, ranges for mesophilic organisms are from 20°C to 45°C, and ranges for thermophilic microorganisms go above 45°C [9].

Most organisms function within a relatively neutral pH range between 6 and 8 [9]. Bacteria can survive in a pH range from 5 to 10, but they thrive between pH values of 6.5 and 8.5. Below 6.5, fungi can become the predominant organism, and poor BOD removal and settling will result [8]. If the high pH is too high, nutrients such as phosphorus will begin to precipitate and become unavailable for use by microorganisms.

High concentrations of toxic inorganic substances, such as cyanide, arsenic, and heavy metal ions in solution, such as copper, lead and zinc, inhibit enzyme formation in the microorganisms, and eventually kill them [8]. Controlling pH levels to permit the formation of metal precipitates minimizes inhibition. This can be an important consideration in the treatment of toxic wastes.

In contrast to naturally occurring compounds, man-made (anthropogenic) organic compounds (e.g., chloroform) are relatively resistant to biodegradation. The reason is that organisms that are naturally present often cannot produce the enzymes necessary to bring about transformation of the original compound to a point at which the resultant intermediates can enter into common metabolic pathways and be completely mineralized [10]. Based on the chemical structure, several guidelines can be used to determine the biodegradability of an organic compound. These guidelines are summarized as follows:

(a) The presence of certain functional groups, e.g., amines, methoxy, sulfonates, and nitro groups, at certain locations in the molecules can make a compound resistant to biodegradation [10].

(b) Materials with unsaturated bonds in their molecules, e.g., alkenes and alkynes, are more amenable to biodegradation than materials exhibiting saturated bonding [1].

(c) Non-aromatic or cyclic aromatic compounds are less resistant than aromatic compounds [1].

(d) Halogenated compounds tend to be resistant [10].

Table 3.3 lists examples of species or groups of organisms and the man-made compounds they have been found to transform. The conditions that prevailed during the observation or experiment (such as aerobic or anaerobic) are also noted.

Biological treatment processes function best when flow, composition, and concentration are relatively constant. The sludge retention time and the organic substrate to microorganism ratio strongly influence the influent organic concentration levels which can be decomposed. For some compounds, the extent of degradation is also strongly influenced by both opportunity for acclimation and by treatment temperature. For example, a treatability study of phenol in an acclimated trickling filter shows that at 20°C 85-100% of phenol is removed, but the removal efficiency drops to only 20% or less at 10°C [11]. Table 3.4 summarizes available data on removal of priority organic pollutants regulated by EPA in full-scale aerobic biological treatment processes.

Table 3.3 Man-made Compounds and Microorganisms That Can Attack Them

Compound Name	Microorganism	Condition
Aliphatics(nonhalogenated)		
Acrylonitrile	Mixed culture of yeast mold, protozoa bacteria, activated sludge	ae
Aliphatics(halogenated)		
Trichloroethane	Marine bacteria	ae
Trichloromethane	Sewage sludge	ae
Trichloroethane, trichloromethane, methyl chloride, chloroethane, dichloroethane, vinylidiene chloride, trichloroethylene, tetrachloroethylene, methylene chloride, dibromochloromethane, bromochloromethane	Soil bacteria	an
Trichloromethanes, trichloroethylene, tetrachloroethylene	Methanogenic (7) culture	an
Trichloroethane, trichloromethane, tetrachloromethane,dichloroethane, dibromochloromethane, 1,1,2,2-tetrachloroethane, bis-(2-chloroisopropyl) ether, bromoform, bromodichloromethane, trichlorofluoromethane, 1,1-dichloroethylene, 1,2-dichloroethylene, 1,3-dichloropropylene,1,2-trans-dichloroethylene	Sewage sludge	ae
Aromatic compounds (nonhalogenated)		
Benzene	Pseudomonas putida (1) Sewage sludge Stabilization pond microbes	ae ae ae
Toluene	Bacillus sp. (1) P. putida	an ae
Nitrobenzene	Stabilization pond microbes	ae
2,4-Dinitrotoluene	Stabilization pond microbes Activated sludge	ae ae
2,6-Dinitrotoluene, di-n-butylphthalate, diphenylhydrazine	Sewage sludge	ae
Creosol	Pseudomonas sp. (1) Aureobasidium pullulans (4)	 ae
Phenol	Pseudomonas, Vibrio, Spirillum; Flavobacterium Chromobacter, Chlamydamonas ulvaensis (2) Phoridium fuveolarum, Scenedesmus basiliensis (2)	ae ae

Table 3.3 Man-made Compounds and Microorganisms That Can Attack Them continued

	Euglena gracilus (2)	
	Corynebacterium sp. (1)	ae
p-Nitrophenol	Rumen microorganisms	an
Aromatic compounds (halogenated)		
1,2, 2,3-, 1,4-Dichlorobenzene, p-, m-, o-chlorobenzoate, 3,4-, 3,5-dichlorobenzoate, 3-methyl benzoate, 4-chlorophenol	Sewage sludge	ae
	Pseudomonas sp. (1), sewage	ae
	Pseudomonas sp. B13 (WR1)	ae
Hexachlorobenzene, trichlorobenzene	Sewage sludge	ae
1,2,3- and 1,2,4-Trichlorobenzene	Soil microbes	ae
Pentachlorophenol	Soil microbes	an
Monochlorophenol, monochlorobenzoate	Nocardia, Mycobacterium (5)	ae
Polycyclic aromatics (nonhalogenated)		
Benzo(a)pyrene	Cunninghamella elegans (4)	ae
	Pseudomonas sp. (1)	ae
	Beijerinckia sp. (1)	ae
Naphthalene	Agnenellum, Oscillatoria (3)	
	Anabaena (3)	ae
	Cunninghamella elegans (4)	ae
	Microcoleus so. (2), Nostoc sp., Coccochlories sp., Aphanocapsa sp., Chlorella sp., Dunaliella sp., Chlamydamonas sp., Cylindriotheca sp., Amphora sp. (2)	ae
	Pseudomonas, Flavobacterium, Alcaligenesae, Corynebacterium, Aeromonas, Flavobacterium (1), Nocardia (5)	
	Stream bacteria	ae
Pyrene	Stabilization pond organisms	ae
Fluoranthene	Sewage sludge	ae
Anthracene	Stream bacteria	ae
Phenanthrene	Beijerinckia (1)	ae
Benzo(a)anthracene (BA)	Cunninghamella elegans	ae
Dibenzanthracene	Activated sludge	ae
Biphenyl	Beijerinckia B8/36	ae
	Oscillatoria sp. (3)	ae
	Pseudomonas putida (1)	ae
Polycyclic aromatic hydrocarbons (halogenated)		
PCBs (mono- and dichlorobiphenyls)	Pseudomonas, Vibrio, Spirillum, Flavobacterium Achromobacter Chromobacter Bacillus (1), Nocardia (5)	ae

Table 3.3 Man-made Compounds and Microorganisms That Can Attack Them continued

4-Chlorobiphenyl, 4,4-dichlorobiphenyl 3,3'- dichlorobiphenyl	Fungi	ae
Pesticides		
Toxaphene	Corynebacterium pyrogenes (1)	an
Heptachlorobornane	Micromonospora chalcea (5)	ae
	Bovine rumen fluid (7)	an
Lindane	Chlorella vulgaris (2)	ae
	Chlamydamonas reinhardtii	ae
	Clostridium sp., Pseudomonas (1)	an
	Soil bacteria	an
	Sewage sludge	an
Dieldrin	Anacystis nidulans (3)	an
	Agmeneloum quardiplicatum (3) Pseudomonas (1)	an
	Rumen fluid (7)	an
	Actinomycetes	an/ae
DDT(1,1'-bis(p-chlorophenyl)-2,2,2-trichloroethane)	Klebsiella pneumoniae (1) E. coli, Aerobacter aerogenes, Pseudomonas, Clostridium, Proteus vulgaries	an
	Sewage	ae
	Soil bacteria	an
	Rumen bacteria (7)	an
	Yeast	ae
	Trichoderma viridae (4)	an
	Fusarium oxysporum (4)	ae
	Mucor alterans (4)	ae
	Cylindrotheca, Closterium (2)	ae
	Dunaliella (2)	ae
	Anaerobic sludge (7)	an
	Nocardia, Streptomyces (5)	ae
	Hydrogenomonas (1)	an/ae
Parathion	Bacillus subtilis, Rhizobium, Chlorella pyrenoidosa, soil bacteria	an
Phorate sulfoxide	Soil bacteria	an
Pentachloronitrobenzene (PCNB)	Aspergillus niger, Fusarium solani, Glomerella congulata, Helminthosporium Victoriae, Myrotheium, Penicillium, Trichoderma viridae (4)	ae
Methoxychlor	Nocardia sp., Streptomyces sp. (5)	ae
	Aerobacter aerogenes (1)	ae/an
Acrolein	Site water (microbes)	ae
Aldrin	Site water (microbes)	ae
	Sewage sludge	an
Endosulfan	Fungi, bacteria, soil actinomycetes	ae
Endrin	Pseudomonas sp., Micrococcus sp., yeast (4)	ae

Table 3.3 Man-made Compounds and Microorganisms That Can Attack Them continued

Compound	Microorganism	
	Sewage sludge	an
Chlordimeform	Chlorella (2), Oscillatoria (3)	ae
Kepone	Treatment lagoon sludge	an
Diuron	Mixed culture of fungi and bacteria; mineralization; single isolates ineffective	ae/an
Nitrosamines		
Dimethylnitrosamine	Rumen organisms (7)	an
	Rhodopseudomonas capsulata (6)	an
Phthalate esters	Micrococcus 12 B	ae
	Sediment-water	an

(1) Bacteria (2) Algae (3) Blue-green algae (4) Fungi (5) Actinomycetes
(6) Photosynthetic bacteria (7) Consortium of anaerobics (8) Reductive dechlorination
(9) Dehydrodichlorination

ae -- aerobic; an -- anaerobic

Table 3.4 Removal of Organic Priority Pollutants by Biological Treatment Systems

Pollutant	Type of Treatment	Influent Level, μg/L	Percent Removal
Acenaphthene	AS-M	0.9	95
	AS-I (10)[a]	15	>84
	AL-I(1)	4.0	0
Benzene	AS-M	8.8	66
	AS-M	7.7	92
	AS	—	90-100
	AL	6-53	(-17)-96
	PACT	160	>99
	AS-I (9)	10,250	>60
	AL-I	46.9	>84
Benzidine	AS-I (1)	4	0
	AL-I	11.9	41
Carbon Tetrachloride	PACT	95	94
	AS-I (2)	250	>98
Chlorobenzene	PACT	1,900	99+
	AS-M	177	99+
	AS-I (6)	15.2	>67
1,2,4-Trichlorobenzene	AS-M	0.6	83
	AS-I (11)	285	>67
Hexachlorobenzene	AS-I (4)	0.75	>47
	AL-I (1)	10	0
1,2-Dichloroethane	AS-M	1.5	65
	AS-M	8.4	85
1,1,1-Trichloroethane	PACT	18	>99
	AS-M	88.4	70
	AS-M	1,791	98
	AS-I (6)	9.2	>74
	AL-I (1)	550	96
1,1-Dichloroethane	AS-M	1.7	6
	AS-M	6.7	92
	AS-I (2)	7.1	>9
1,1,2-Trichloroethane	AS-I (1)	11	>9
1,1,2,2-Tetrachloroethane	AS-I (2)	12.8	>22
Bis(chloromethyl) Ether	AS-I (1)	59	83
Bis (2-chloroethyl) Ether	AS-M	0.12	92
	AS-I (1)	19	>47
Naphthalene	AS[b]	2,000	100
	AS-I (1)	2	50
	TF-I (1)	2	0
	AL-I (1)	19	>47
2,4,6-Trichlorophenol	AS-I (10)	703	36
	AL-I (1)	1,000	99
Chloroform	AL	425-2,645	(-1)-86
	AS-M	4.2	5
	AS-M	17.1	98
	AS-M	37.5	49
	AS-I (16)	33	>61
	TF-I (1)	19	0
	AL-I (3)	531	>36
2-Chlorophenol	AS-I (2)	10	46
Dichlorobenzene	PACT	720	35

Table 3.4 Removal of Organic Priority Pollutants by Biological Treatment Systems continued

1,2-Dichlorobenzene	AS-M	1.8	28
	AS-M	4.8	28
	AS-M	43.9	97
	AS-I (12)	28.8	>74
	AL-I (1)	250	>96
1,3-Dichlorobenzene	AS-M	3.1	86
1,4-Dichlorobenzene	AS-I (8)	30	>82
	AL-I (1)	53	>81
1,1-Dichloroethylene	AS-M	43.2	97
2,4-Dichlorophenol	AS-I (2)	13.3	>25
1,2-Dichloropropane	AS-I (2)	16	>67
2,4-Dimethylphenol	AS-I (3)	39.7	>32
2,4-Dinitrotoluene	PACT	2,000	75
	AL-I (1)	3	0
2,6-Dinitrotoluene	PACT	1,900	76
	AS-I (1)	390	0
	AL-I (1)	12	>83
1,1-Diphenylhydrazine	AS[b]	~2,000	100
	AS-I (1)	340	0
	AL-I (1)	14	0
Ethylbenzene	PACT	29	78
	AS-M	148	99
	AS-I (24)	882	83
	AL-I (3)	45	>78
Fluoranthene	AS-M	0.6	92
	AS-I (1)	2	0
	AL-I (1)	2	0
4-Bromophenyl Phenyl Ether	AS-I (1)	360	95
Bis (2-chloroisopropyl) Ether	AS-M	0.16	62
	AL-I (1)	2	0
Bis (2-chloroethoxy) Methane	AL-I (1)	25	>60
Methylene Chloride	AS-M	9.0	49
	AS-M	293	0
	AS-M	647	72
	AS-I (5)	144	34
	TF-I (1)	1	0
	AL-I (3)	1,114	65
Methyl Chloride	AL-I (1)	56	>91
Bromoform	PACT	910	89
	AS-I (1)	3	0
Dichlorobromomethane	PACT	54	>99
	AS-I	5.8	0
Trichlorofluoromethane	PACT	920	99
	AS-I (5)	556	19
	TF-I (1)	48	>79
Chlorodibromomethane	PACT	81	>99
Isophorone	AS-I (2)	10	>0
	AL-I (1)	3.0	33
Naphthalene	AS-M	1.3	35
	AS-M	10.9	62
	AS-I (26)	50	>64
	TF-I (1)	55	0
	AL-I (2)	14.1	>29
	TL-I (1)	56	56
Nitrobenzene	AL-I (1)	10	0
2-Nitrophenol	AS-I (1)	40	>99
4-Nitrophenol	PACT	56	>99
	AS-I (1)	90	>99

Table 3.4 Removal of Organic Priority Pollutants by Biological Treatment Systems continued

	AL-I (1)	13	23
2,6-Dinitro-o-cresol	PACT	11	99
Nitrosodimethylamine	AS	~2,000	>95
N-Nitrosodiphenyl Amine	AS-I (2)	5.3	>84
	AL-I (1)	3	67
N-Nitrosodi-n-propylamine	AS-M	0.5	88
	AS-M	6.7	99
	AS-I (2)	11	0
Pentachlorophenol	AS-I (15)	5,333	>70
	TF-I (1)	3	0
	AL-I (1)	34	>71
Phenol	TF	105	91
	AS	105	>99
	PACT	440	96
	AS-I (30)	335	>77
	TF-I (1)	37	0
	AL-I (3)	25.4	>41
Bis (2-ethylhexyl) Phthalate	PACT	2.4	67
	AS-M	1.3	31
	AS-M	50	78
	AS-I (38)	102	37
	TF-I (1)	35	83
	AL-I (5)	37	>70
	TL-I (2)	26	>58
Butyl Benzyl Phthalate	AS-M	18	91
	AS-I (1)	11	0
	AL-I (1)	6	0
Di-n-butyl Phthalate	AS-M	5.6	68
	AS-M	13	62
	AS-I (9)	23	>60
	TF-I (1)	15	25
	AL-I (1)	1	0
Di-n-octyl Phthalate	AS-I (1)	5,000	0
Diethyl Phthalate	AS-M	6.6	98
	AS-M	1.4	79
	AS-I (17)	15	56
	TF-I (1)	140	0
	AL-I (1)	4	0
Dimethyl Phthalate	AS-I (9)	60	60
	AL-I (1)	8	25
Benzo(a)pyrene	AL-I(1)	3	33
Acenaphthylene	AS-M	0.2	80
	AL-I (1)	5	0
Anthracene	AS-I (7)	8.5	>60
Fluorene	AS-M	1.7	94
	AS-I (2)	2	>99
Phenanthrene	AS-M	3.2	91
	AL-I (1)	3	0
Pyrene	AS-I (5)	2.3	16
	AL-I (1)	3	67
Tetrachloroethylene	PACT	62	88
	AS-M	6.4	0
	AS-M	49	80
	AS-M	560	>99
	AS-I (11)	25.6	>75
	AL-I (1)	25	>60
Toluene	PACT	680	>99
	AS-M	70	>99

Table 3.4 Removal of Organic Priority Pollutants by Biological Treatment Systems continued

	AS-M	112	99
	AS-I (31)	119	52
	AL-I (3)	143	>93
Trichloroethylene	AL	10-120	(-23)-73
	PACT	60	90
	AS-I (12)	31.3	68
	TF-I (1)	1	0
Aldrin	AS+MMF[b]	---	100
Diedrin	AS+MMF[b]	---	72
DDT	AS+MMF[b]	---	100
DDE	AS+MMF[b]	---	39
DDD	AS+MMF[b]	---	100
	AS-M	0.3	37
	AS-M	0.17	17
Heptachlor	AS-I (1)	6.3	76
Lindane	AS+MMF[b]	---	0

AS = Activated Sludge; M = Municipal Facility; I = Industrial Facility; AL = Aerated Lagoon TF = Trickling Filter; MMF = Mixed Media Filtration; PACT = Activated Sludge with Powdered Activated Carbon added; TL = Tertiary Polishing Lagoon.

[a] Numbers in parentheses represent number of treatment plants for which data were averaged.

[b] Pilot scale data.

Source: [9]

3.2. DESCRIPTION OF BIOLOGICAL TREATMENT PROCESS

Biological treatment processes available for wastewater treatment can be classified according to the nature of their biological growth. Those in which the active biomass is suspended as free organisms or microbial aggregates are referred to as suspended growth treatment systems, whereas those in which growth occurs on or within a solid media are referred to as supported growth systems. Both of these systems can be designed to operate under aerobic or anaerobic conditions. Several widely used aerobic processes in both suspended growth and supported growth configurations, and an anaerobic suspended growth process (anaerobic contact process) will be described here including process descriptions, suitability for treating hazardous chemicals, and cost estimates.

3.2.1. Aerobic Suspended Growth Process

In suspended growth treatment, microorganisms are flocculated in a liquid medium and oxygen is supplied by aerators which provide good air/water contact. Organics are usually removed as nutrients in the course of microbial metabolism. The stable floc formed can then be separated from the treated effluent stream by sedimentation.

3.2.1.a. Process Description -- Activated Sludge Process

The activated sludge process is a typical type of suspended growth biological treatment system and probably the most widely used biological process for the treatment of organic and industrial waste waters. However, it can only treat aqueous organic wastestreams having less than 1% suspended solid content [12], and can not tolerate shock loadings of concentrated organics. Therefore, the wastestream entering this process will usually have passed through a pretreatment process which includes a clarifier (primary clarifier) and an equalization basin. The primary clarifier is used for removal of grit, oily and fatty material and gross solid material, while the equalization basin is used to dampen wastewater flow variations and to provide more uniform organic loading to the activated sludge system.

After passing through the pretreatment process the clarified wastestream with a uniform organic concentration is then brought into contact with a mixed microbial population in the form of a flocculent suspension in an aerated and agitated vessel (aeration basin). Suspended and colloidal material is removed from the waste water by adsorption and agglomeration onto the microbial flocs. This material and dissolved nutrients are then broken down more slowly by microbial metabolism. Sufficient residence time in the vessel is provided to achieve the desired degree of treatment. As flow leaves the vessel, the flocculent microbial mass (sludge) is separated from the treated waste water by gravity settling in a separate vessel (secondary clarifier). Most of the settled sludge is returned to the aeration basin to maintain the sludge concentration at the level needed for effective treatment and to act as a microbial inoculum. The excess is wasted and is removed for disposal after passing through a post-treatment process (e.g. thickening and dewatering processes). A simple flow diagram of activated sludge process is shown in Fig. 3.2.

For the treatment of industrial waste water, supplemental nutrient sources are often needed to provide sufficient nitrogen and phosphorus. In most cases, nitrogen is added as ammonia and phosphorus as phosphoric acid. A proper pH range (6-8) and a sufficient dissolved oxygen concentration (a minimum of 1-2 mg/l) must also be maintained in the aeration basin to support a healthy and active system [8].

The aeration basin hydraulic retention time (HRT) and sludge residence time (SRT) are important operational factors. HRT is defined as the ratio of the volume of aeration tank to the influent liquid flow rate, and SRT is the total amount of sludge in the system divided by the rate of sludge leaving the system as waste. Sufficient time must be provided to allow the bacteria to assimilate the organic material in the waste water. The HRT is usually from 6 to 24 hours and SRT is from 4 to 10 days for the activated sludge process [12]. The optimum operating temperature is in the range of 25 to 32°C [8].

Table 3.5 lists the BOD loadings and removal efficiency of the activated sludge process at different operating conditions. Extended aeration which is also called "low-rate" treatment operates at very high sludge residence times (up to 24 days). The high residence time and the low BOD

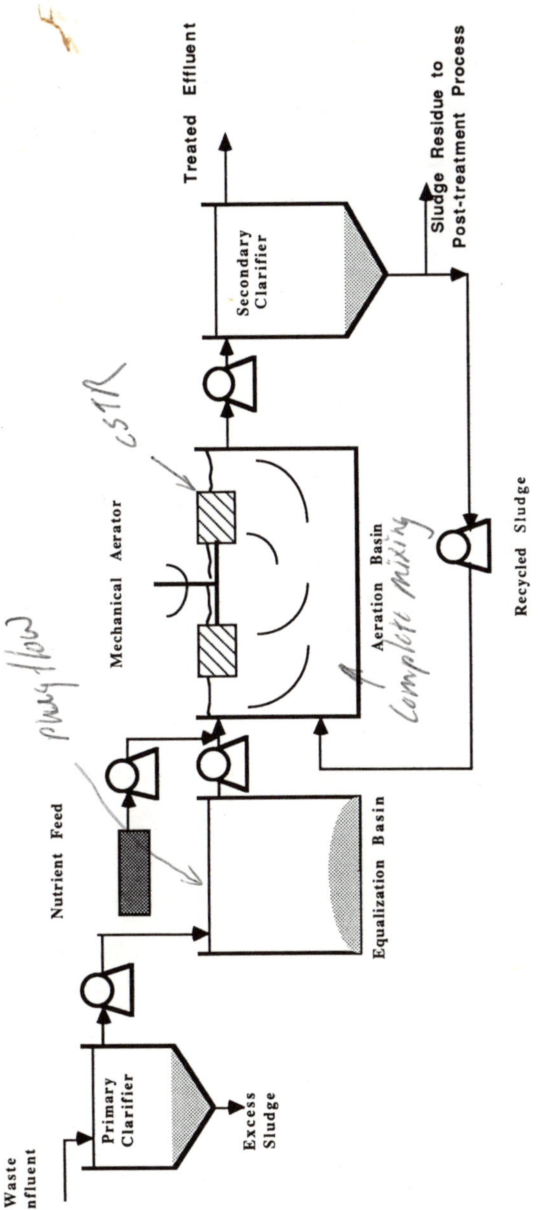

Figure 3.2. Flow Diagram of the Activated Sludge Process

Table 3.5 Activated Sludge Loadings

Type of plant	BOD load (Kg/Kg MLSS*)	Efficiency of BOD removal (%)
Extended aeration	0.02-0.06	95
Conventional load	0.2-0.5	90-95
High rate	1.5-2.0	60-70

*Mixed liquor suspended solids
Source: [2,27]

loading in this treatment suggest that much of the population of bacteria is well into the endogenous respiration phase where sludge breakdown exceeds sludge growth. The sludge production rate may be as low as 0.2 to 0.3 kg solids per kg BOD removed. At the other end of the spectrum, in the high-rate treatment process, the BOD loading levels are several times higher than those in a conventional process and the sludge residence time is short (<0.5 day), giving a rapid, partial treatment. This process is often used as a preliminary stage preceding a more complete treatment process.

Although organisms present in activated sludge systems range from viruses to multicellular organisms, the predominant and most active are heterotrophic, and to lesser extent, autotrophic bacteria, which are both aggregated in the sludge flocs and dispersed in the liquid [13]. Heterotrophic bacteria utilize organic material as a source of both carbon and energy, while autotrophic bacteria generally depend on the oxidation of mineral compounds for energy requirements and utilize carbon dioxide as a carbon source. These bacteria are capable of performing hydrolysis and oxidation reactions.

Complex hydrocarbons are oxidized to lower molecular weights by oxygenase enzymes which incorporate oxygen directed into the long chain or cyclic hydrocarbon molecule. Polysaccharides, fats, and proteins are degraded from their polymeric state to monomeric units via hydrolysis [14]. The end-products, i.e., alcohols and acids, from those reactions will enter the microorganism and be metabolized by oxidation reactions catalyzed by endo-enzymes. The oxidation follows the chemical sequence of: alcohols oxidized to aldehydes and then to acids. A portion of the acids are oxidized to carbon dioxide and water to obtain the necessary energy to use remaining acids for cell growth [9].

Generally, the activated sludge process is readily capable of decomposing alcohols, aldehydes, fatty acids, alkanes, alkenes, cycloalkenes and aromatics [14]. Other compounds such as isoalkanes and halogenated hydrocarbons are more resistant to microbial decomposition. Therefore the degree of treatment and the rate of decomposition are dependent upon the acclimated biomass in the activated sludge system. However, only dilute aqueous wastes can normally be treated, and most hazardous organic wastes are toxic or inhibitory to the process except at very low concentrations. Therefore, treatment of hazardous wastes by this process is often most practical where the aqueous waste can be mixed with a more readily biodegradable wastewater stream [12].

Dissolved metal ions and fine metal particles produce an adverse effect on microbial metabolism by binding at the enzyme-active site or causing conformational changes in the enzyme with the activated sludge process. Normally, microorganisms can tolerate only a few milligrams per liter or less of heavy metals. The toxicity-threshold concentrations of various metals and inorganics to mixed activated sludge cultures are listed in Table 3.6. Heavy metals may be kept insoluble by the addition of ferrous sulfate to encourage sulfide precipitation and light metal cations may be detoxified by encouraging formation of carbonates and bicarbonates. In addition to biodegradation, organic materials may be removed by air-stripping, and/or sorption to the sludge. Results of bench scale tests to determine the relative degree of air-stripping and biodegradation for selected compounds are shown in Table 3.7. Stripping problems can be eliminated by modifications to the process, such as pre-dissolving high purity oxygen into the influent wastestream (Dorr-Oliver Biological Systems Company) [5] or using a closed pressurized aerator with pure oxygen (Union Carbide Chemical Company) [1]. Heavy metals such as cadmium, chromium, copper, lead and zinc, and high molecular weight refractory organic compounds

Table 3.6 Toxicity-Threshold Concentrations for Various Metals and Inorganics in Activated Sludge Process

Component	Concentration (mg/l)
Silver	<0.03
Vanadium	10.0
Zinc	2.0
Nickel	1.0-2.5
Chromium, $^{+6}$	10.0
Chromium, $^{+3}$	50.0
Lead	0.1
Iron (Ferric)	15.0
Copper	1.0-10.0
Cadmium	1.0
Arsenic	0.1
Boron	0.05-100
Cyanide	0.1-5
Manganese	10
Mercury	0.1-5.0
Ammonia	480

Source: [9,13]

Table 3.7 Biodegradation and Air-stripping Removal Efficiency of Various Organic Compounds in Activated Sludge Process

Compound	Overall(%)	Biodegradation(%)	Air stripping(%)
Tetrachloroethane	99.2	6.2	93
Nitrobenzene	99.1	99.1	0
2,4-Dichlorophenol	98.8	98.8	0
Acrolein	96.7	96.7	0
Acrylonitrile	97.5	97.5	0
1,2-Dichloropropane	99.2	0.2	99
Methylene chloride	99.3	94.3	5
Ethyl acetate	99.4	82.4	17
Benzene	99.2	84.2	15
1,2-Dichloroethane	99.3	1.3	98
Phenol	99.6	99.6	0
1,2-Dichlorobenzene	98.2	74.2	24

Source: [26]

including PCBs, cyanide, chlordane, DDT, dieldrin and lindane have been found to concentrate in the sludge stream [9]. The effects of these removal mechanisms on the environment are important, and have to be considered in the design of the activated sludge process.

3.2.1.b. Cost Estimation

The investment for the activated sludge process includes the costs of the pretreatment process (i.e., primary clarifier and equalization basin), aerator and secondary clarifier. The cost for each of these individual process unit depends on the specific design requirements such as solid concentration for clarifier, and solid and liquid retention time for aerator. Because of this, the actual vendor costs are difficult to obtain. Therefore, literature values are used throughout this study. The literature cost data for those unit processes are calculated based on the following design assumptions [15]:

(1) Primary clarifier:

circular shape

sludge concentration = 4% solid

(2) Equalization basin

retention time = 1 day

mixing requirements = 20 to 40 hp/Mgal

(3) Aerator

volumetric organic loading = 32 lb BOD/(day)(10^3 ft^3)

(1.1 lb O_2 supplied)/(lb BOD removed)]

oxygen transfer rate = 1.8 lb/(hp-hr)

hydraulic retention time = 6 hr

F/M (Food to Microorganism ratio) = 0.25 lb BOD/(day)(lb MLVSS)

MLVSS (Mixed Liquor Volatile Suspended Solid) = 2100 mg/l

(4) Secondary clarifier

circular shape

surface overflow rate = 800 gal/(day)(ft^2)

sludge concentration = 1% solids

The construction costs for flows ranging from 0.1 to 1 Mgal/day for each unit process are shown in Fig. 3.3. The prices have been updated to 1987 using the published CE plant cost index [16]. To obtain the overall capital cost for the activated sludge process, the non-component and non-construction cost must also be considered. The non-component cost, which includes piping, electrical, instrumentation and site preparation, is about 28% of the sum of the unit process construction costs [15]. The non-construction cost which includes engineering and construction supervision and contingencies is about 30% of the sum of the above two costs (construction and non-component cost). Assuming an interest rate of 10% and 25 years of annual installments, the cost of annual capital amortizations can be calculated from the following capital recovery equation using the capital cost shown in Fig. 3.4.

$$\text{Annual Capital Cost} = \text{Capital Cost} * [\, i\,(i+1)^n / (i+1)^n - 1\,] \tag{3.1}$$

where i is the interest rate (0.1) and n is the number of years (25). The results are also shown in Fig. 3.4.

The operation and maintenance (O&M) costs are for operation and maintenance labor, materials, chemicals, and electrical power. The annual O&M costs for each of the unit processes are shown in Fig. 3.5 [15]. Similar to the construction cost, the annual O&M cost for the activated sludge process is the sum of the O&M cost for each of the unit processes. This result is shown in Fig. 3.5. The annual operating costs, which are defined as the sum of the annual capital cost and annual O&M costs, are shown in Fig. 3.6.

3.2.2. Aerobic Supported Growth Systems

Supported growth biological treatment is based on developing a biological growth on a solid surface and then passing organic wastes over the surface or passing the surface through the

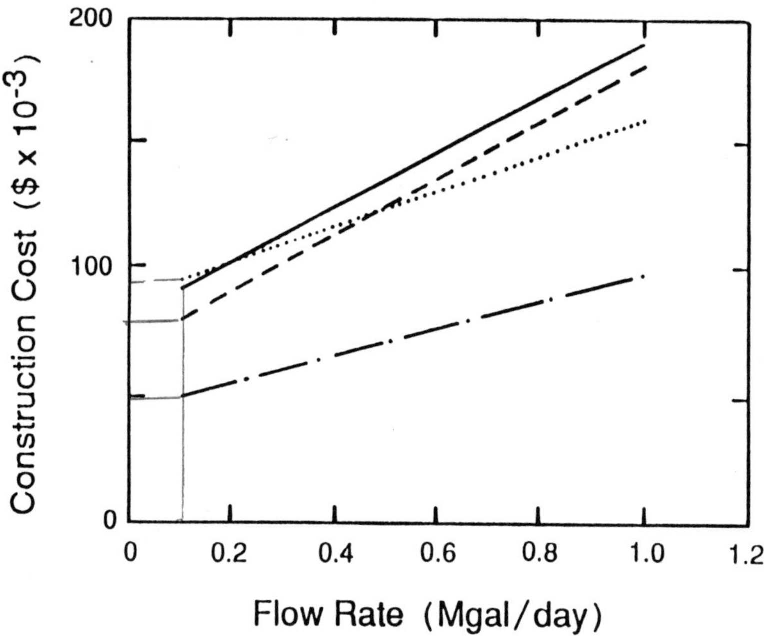

Figure 3.3. Construction Costs for Aerator, Pretreatment and Post-treatment Units.
—·— Primary Clarifier; — — - Secondary Clarifier;
——— Aereator; .... Equalization Basin.

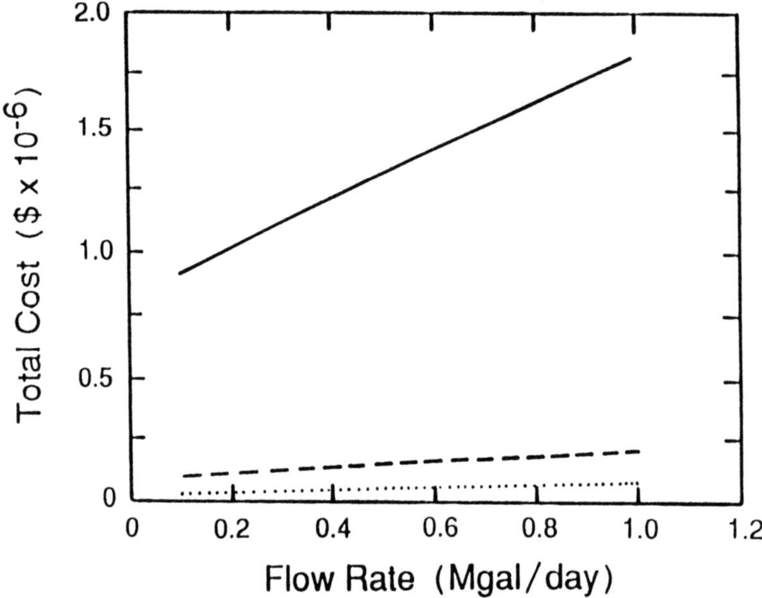

Figure 3.4. Costs for the Activated Sludge Process.
Capital (——) , Annual Capital (— —), and Annual O & M (·········)

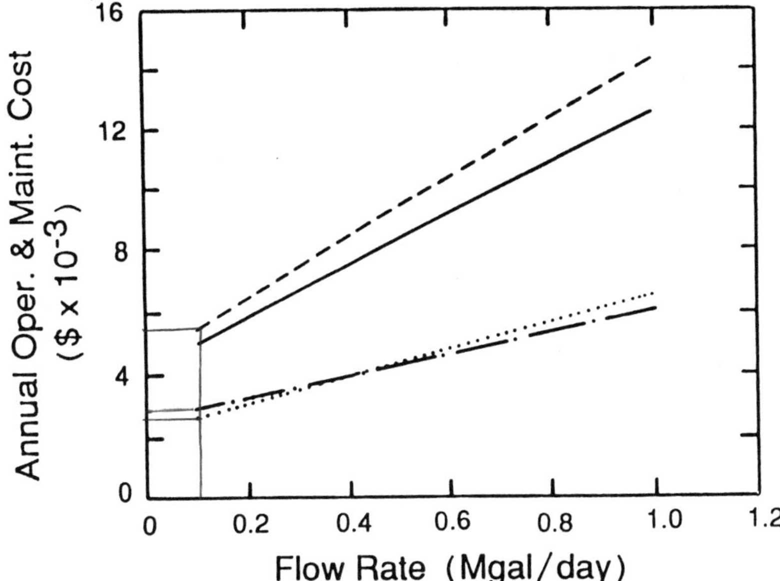

Figure 3.5. Annual O & M Costs for Aerator, Pretreatment and Post-treatment Units. (·········) Secondary Clarifier; (—— · ——) Primary Clarifier; (———) Aerator; (— —) Equalization Basin.

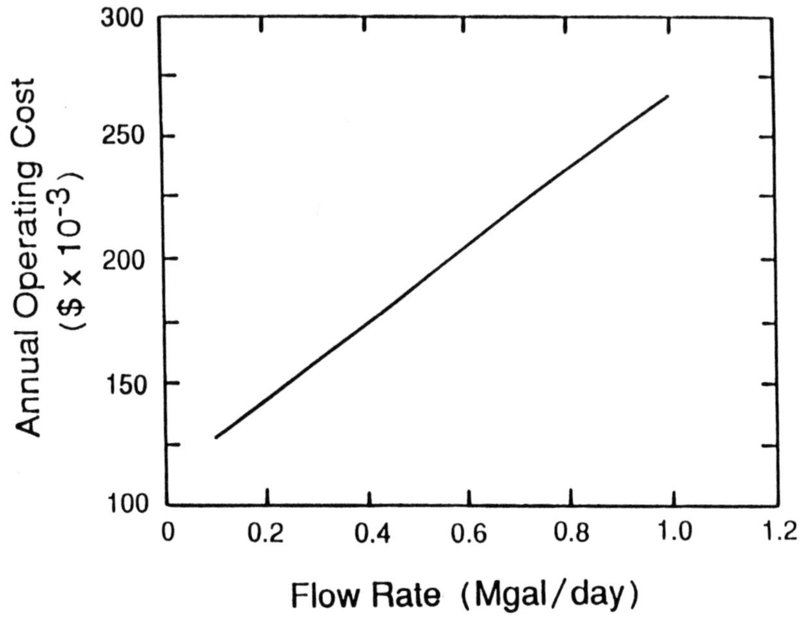

Figure 3.6. Annual Operating Cost for Activated Sludge Process.

wastes. The microbial culture attached to the solid surface is able to decompose organic matter in the wastestream. Two distinct types of attached-growth biological treatment processes, trickling filter (stationary medium) and rotating biological contactor (moving medium), will be discussed here.

3.2.2.a. Process Description

Trickling Filter

The trickling filter process brings the wastewater in contact with aerobic microorganisms, by trickling the wastewater through a fixed bed of rocks or synthetic media that support the microorganisms. The wastewater is distributed over the top of the bed through fixed sprays or moving arrays of spargers. Aeration of the bed takes place by natural convection due to the difference in temperature between the air in the packing media and that of surrounding atmosphere, although forced convection is sometimes used [2].

The wetted surface of the packing medium develops a film of microbial culture, and the wastewater flows over the packing surfaces in a thin layer which is in contact with the microbial culture on one side and the atmosphere in the interstitial spaces on the other (Fig. 3.7). Oxygen dissolves at the surface of the moving liquid layer and is transferred across the liquid layer to the microbial film. The oxygen and nutrients from the liquid diffuse into the microbial film to be metabolized by the microbial population. Suspended and colloidal materials in the wastewater are also agglomerated and adsorbed into the microbial film. The microbial film remains aerobic primarily at its surface where air and water interface with the cells. The underlying portion, adjacent to the media, may become anaerobic. Thus, the biological treatment encompasses both aerobic and anaerobic degradation.

Gross solids in wastestreams can cause blockages in the packed bed if they become trapped in the packing medium. Therefore, similar to the activated sludge process, an appropriate physical pretreatment process has to be installed before the filter. Periodically, the microbial film coating sloughs off the media. This sloughing off may occur for several reasons. As the microorganisms

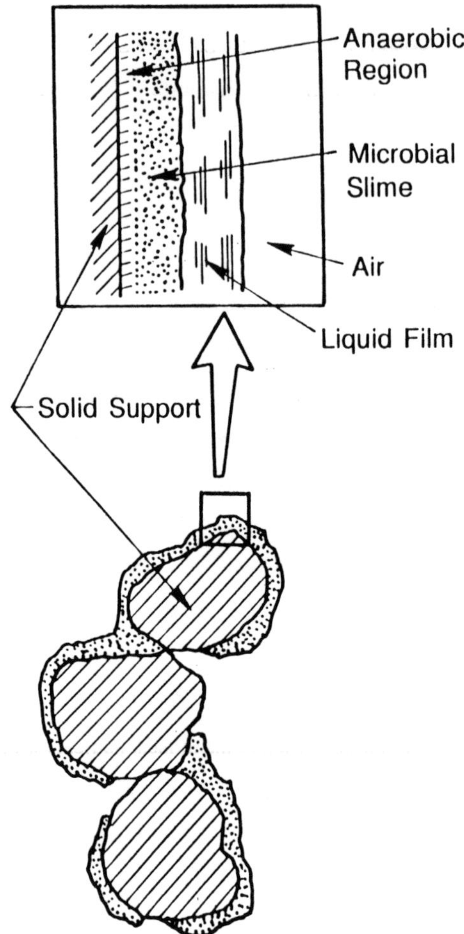

Fig. 3.7. Typical Microbial Film Attached to a Solid Supporting Medium.

grow and reproduce, the film coating may become too heavy to remain attached to the media. Also, as the supply of substrate to the underlying microorganisms becomes limited, endogenous respiration may occur. During endogenous respiration, the microorganisms become more dense and their adhesive polysaccharide cell coating diminishes. The detached film (humus) carried out of the system in the filter effluent is separated by settling in a secondary clarifier.

Recirculating of part of the treated effluent to mix with the influent wastewater feed is necessary for increasing the contact time of the waste with the attached biological mass which increases the removal efficiency of the wastes, and for balancing the hydraulic loading with the nutrient loading which allows the system to treat the wastewaters with very high organic concentrations. The recirculated treated effluent is usually taken from the secondary clarifier output, rather than directly from the filter effluent, as the solid-free liquid reduces the risk of clogging the packing with re-introduced humus [2]. A flow diagram of the trickling filter process is provided in Fig. 3.8.

Trickling filter biological treatment can be classified broadly as "high-rate" or "low-rate" according to the hydraulic loading or nutrient loading rate of the system [17]. In general terms, low-rate processes remove a high percentage of the influent nutrients, but the rate of removal is low, in terms of the mass of nutrients removed per unit volume of the system. Conversely, high-rate treatments remove a lower proportion of the influent nutrients with a high rate of removal. Consequently, high-rate processes are sometimes called "roughing" treatments and low rate processes "polishing" treatments. Some biological treatment processes use two filters in series. The first filter in this so called "double filtration" is operated at a high loading rate and the second at low-rate loading. About 70% of the nutrients in the influent are removed in the first stage, and after removal of humus in a separation stage, the liquid passes to the second stage where nearly all the remaining nutrients are removed. In order to prevent the first filter from becoming blocked by biomass due to the high loading rate, an alternating double filtration (ADF) scheme was developed [4]. The order of the filters is reversed before the biomass in the first filter prevents the free flow of waste and air through it. The original primary filter then receives treated effluent which has

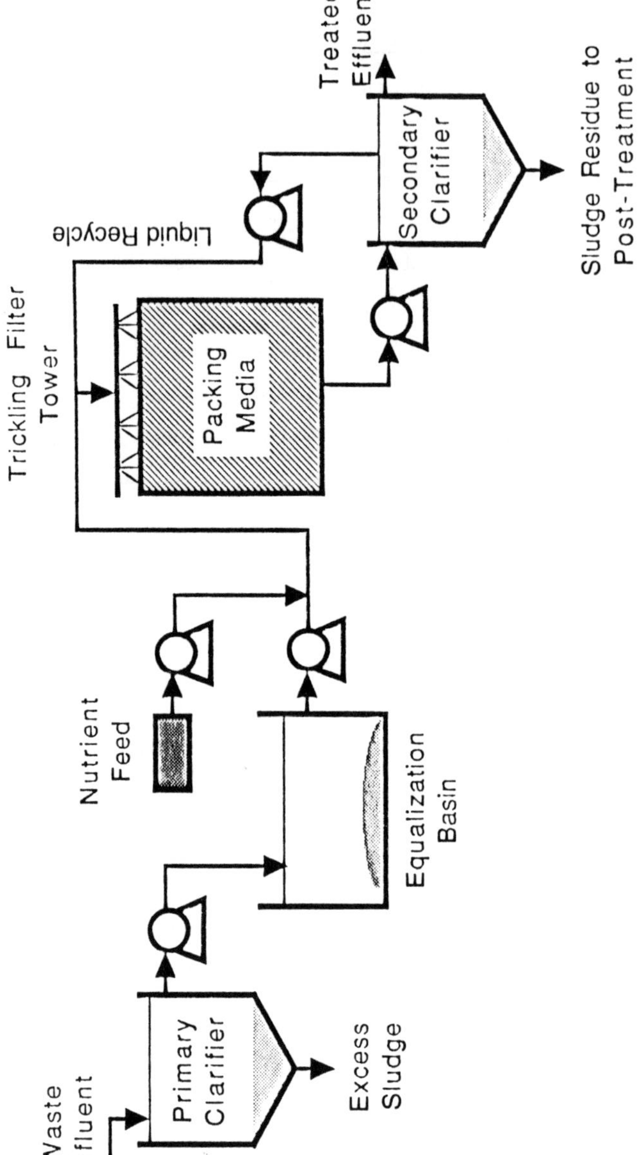

Figure 3.8. Flow Diagram of the Trickling Filter Process.

passed through the new primary filter. This results in the rapid autolysis and consumption of the starved biomass. ADF enables filters to be operated at more than twice the normal load [6].

A wide variety of packing materials are available for trickling filters. The types of packing in common use may conveniently be classified in two groups, mineral and plastic media. The mineral media such as stones and metallurgical coke have been in use for a very long time principally in sewage treatment plants. The disadvantages of those packing media are their weight, which effectively limits the depth of the beds used, and their low voidage (e.g. 40 to 55% void fraction for 40 to 50 mm stones), which permits the packed bed to be more easily blocked by the microbial film. These problems have been overcome by the development of plastic packing media. The resultant packings have very high voidage, usually about 90%, with wide interstitial spaces in the packing, and are about one-tenth of the weight of mineral media in operation [2]. The BOD loading rate and the removal efficiency of BOD for these treatments are listed in Table 3.8.

Like the activated sludge, the trickling filter can only treat the aqueous organic wastestream with low suspended solids concentration (<1%). Because the short residence time of the wastewater in the process, it can accept sudden changes in hydraulic and organic loadings. The system has been used to level loads and reduce biodegradable organic concentrations prior to other biological treatment processes [9]. Extremes in loading should, however, be avoided so that the biomass does not slough off the filter. Hydraulic loading rates for trickling filters are generally less than 0.02 gpm/m^2(superficial bed area); however, loadings up to 1.6 gpm/m^2 have been employed [18]. The hydraulic loading affects the hydraulic residence time (HRT) of the liquid trickling through the packing. Within the usual range of operation, the retention time decreases with increasing hydraulic loading.

Trickling filters are generally applicable to the treatment of the same types of hazardous wastes that are treatable by activated sludge. Because of the relatively short residence time of wastewater in contact with microorganisms, however, the percentage removal of organics is not as high as in activated sludge treatment. Greater removals are achieved as the depth of media and the recycle ratio are increased. Trickling filters are reported to have successfully handled the following

Table 3.8 Trickling Filter Loadings

Type of plant	BOD load (Kg/m^3 media)	Efficiency of BOD removal (%)
Low-rate filtration	0.05-0.1	95
Double filtration , ADF	0.1-0.5	85-90
High-rate filtration	2.0-6.0	50-70

Source: [27]

waste constituents: acetaldehyde, acetic acid, acetone, acrolein, alcohols, benzene, butadiene, chlorinated hydrocarbons, cyanides, epichlorohydrin, formaldehyde, formic acid, ketones, monoethanolamine, propylene dichloride, and resins [9].

The advantage of the trickling filter process compared to other wastewater treatment systems is that no power is consumed in agitation or aeration for the creation of the gas-liquid contact area. Power is consumed only in transferring liquid to and from the unit and in distributing it over the packed bed, so that the operating cost is low.

Rotating Biological Contactor

In the rotating biological contactor (RBC), a microbial film is built up on a partly submerged support medium which rotates slowly on a horizontal axis in a tank through which the wastewater flows. The microbial film is thus exposed successively to the nutrients in the wastewater and to air as the medium rotates. This motion maintains the biomass in an aerobic condition. The support medium is available in several configurations, such as discs, lattice construction, or a container of plastic balls. The medium is rotated at a speed of about 1 to 7 revolutions per minute using either a mechanical or air-induced drive system [2]. The actual motion of the biological surface is at right angles to the liquid path at most points. This generates turbulence at the solid-liquid interface which permits high mass transfer of nutrients and oxygen into the biological film and enhances sloughing of the excess film into the tank [19]. Figure 3.9 provides a flow diagram of the RBC.

The hydraulic retention time is similar to that of trickling filter (e.g. about 20 minutes for a three-stage RBC with a 50-disk array in each stage) [19]. However, the RBC requires only 10% of the ground area that is needed for the trickling filter. Because of the relatively low HRT it has good resistance to sudden changes in operating conditions. In addition, the RBC process offers several advantages over other types of biological treatment process such as operational simplicity, low power requirement, and high treatment efficiency [19].

3.2.2.b. Cost Estimation

The construction cost data for the trickling filter and RBC obtained from the literature are based on the following design criteria [15]:

Trickling Filter

- plastic media
- bed depth = 21 ft
- hydraulic loading rate = 0.65 gal/(min)(ft^2)
- recirculation ratio = 2:1
- organic loading rate = 20 lb BOD/(10^3 ft^3)(day)

RBC

- hydraulic loading rate = 1.0 gal/(day)(ft^2)

The construction cost data for the trickling filter and the RBC have been adjusted to 1987 prices using published CE plant cost index [16]. The costs for a range of flow rates are shown in Fig. 3.10. The construction cost for the RBC is always higher than that for trickling filter at the same influent flow rate. The annual O&M costs for the filter and RBC units are shown in Fig. 3.11.

The pretreatment process (i.e., primary clarifier and equalization basin) and the post-treatment process (secondary clarifier) for these two aerobic supported growth systems are

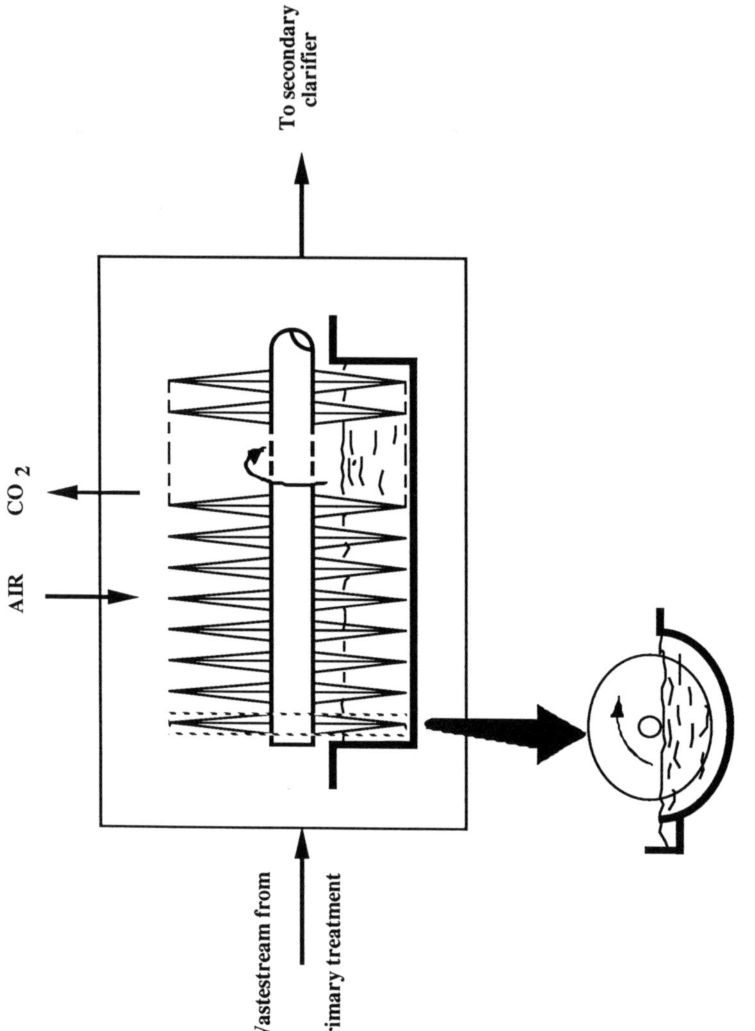

Figure 3.9. Flow Diagram of the RBC.

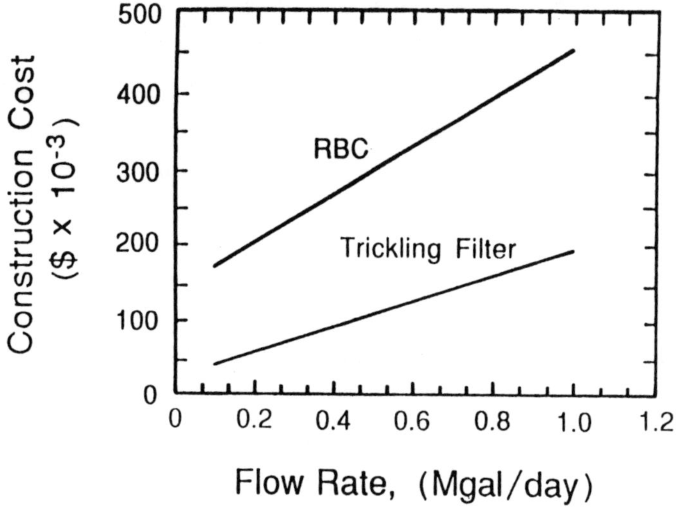

Figure 3.10. Construction Costs for RBC and Trickling Filter Processes.

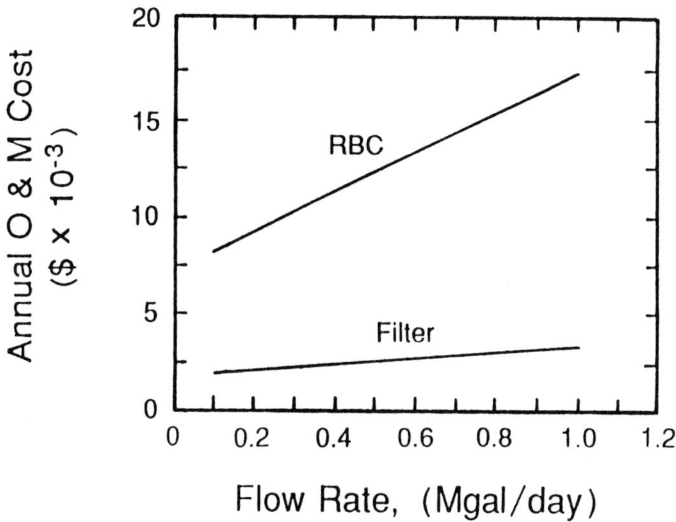

Figure 3.11. Annual O&M Costs for RBC and Trickling Filter Processes.

assumed to be the same as those in the activated sludge process. By using the unit construction prices and unit annual O&M costs for the pretreatment, post-treatment and process units shown in Figs. 3.3, 3.5, 3.10 and 3.11, and following the same procedures and constraints discussed in Section 3.2.1.b., the capital cost, annual capital cost and the annual O&M cost of these processes are calculated. The results at different flow rates are plotted in Figs. 3.12, 3.13 and 3.14 respectively. The results for the annual operating cost are shown in Fig. 3.15.

3.2.3. Anaerobic Contact Process

Anaerobic digestion has traditionally been used as a supporting process in the field of biological treatment. Organic solids and microbial sludges from primary and secondary clarification are processed anaerobically to reduce their volume and improve their stability. However, the advantages of the anaerobic process, lower energy consumption and lower solids production compared to aerobic processes, have resulted in the anaerobic process becoming an important treatment process for industrial waste [20]. Table 3.9 summarizes the advantages and disadvantages of anaerobic processes.

3.2.3.a. Process Description

The anaerobic contact process, similar to the activated sludge process, consists of a high-rate digestor followed by a clarifier. The high-rate anaerobic digestor is a closed tank with provisions for mixing. Mixing may be accomplished by mechanical agitators, gas circulation or pumping. The SRT of high-rate anaerobic digestors is at least three days and averages 14-16 days depending on the operating temperature [9]. Sludge is gravity settled in the clarifier and recycled back to the digestor so that a high level of microorganisms is maintained within the digestor. The HRT for the anaerobic contact process is between 6-12 hours [9]. Unlike most aerobic processes, this process can accept up to 0.5 lb of suspended solids per cubic foot of digestion space per day [9] and COD loading rate of 1 to 6 kg/(m^3)(day) [21]. A schematic diagram of this process is shown in Fig. 3.16.

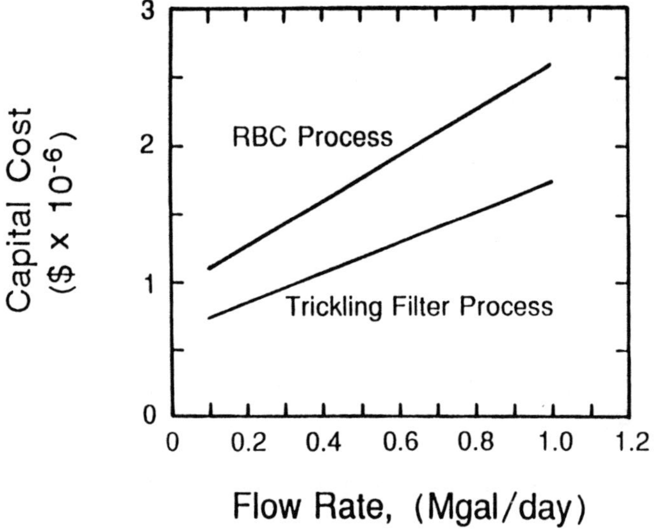

Figure 3.12. Capital Costs for RBC and Trickling Filter Processes.

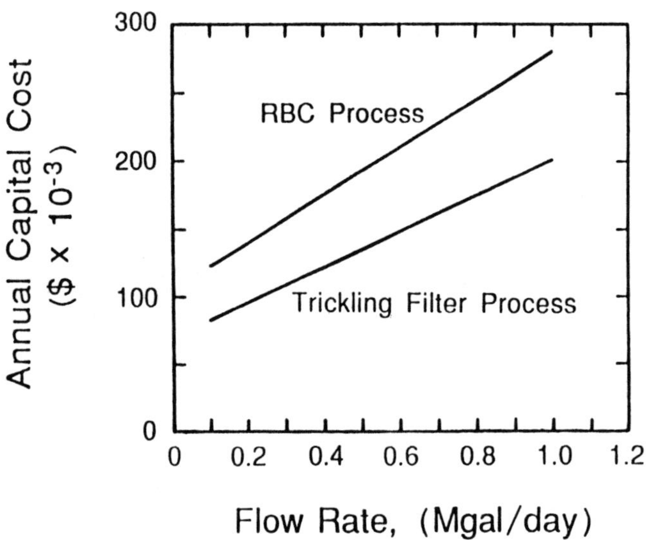

Figure 3.13. Annual Capital Costs for RBC and Trickling Filter Processes.

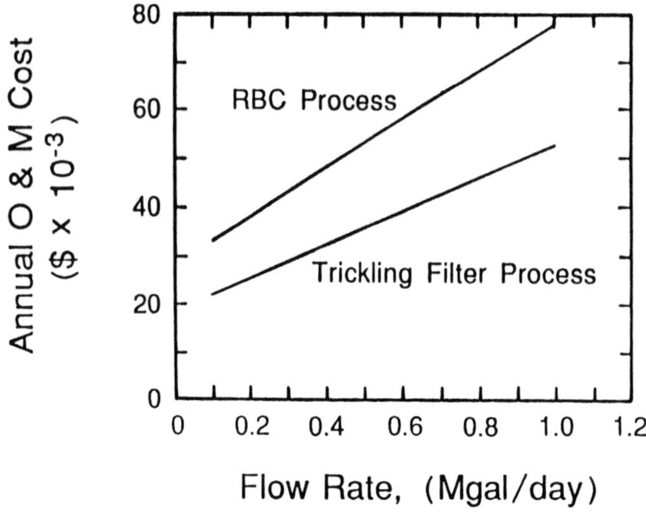

Figure 3.14. Annual O&M Costs for RBC and Trickling Filter Processes.

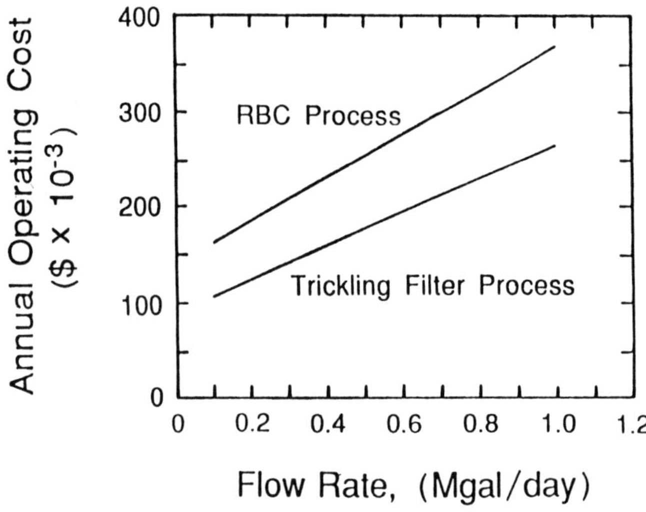

Figure 3.15. Annual Operating Costs for RBC and Trickling Filter Processes.

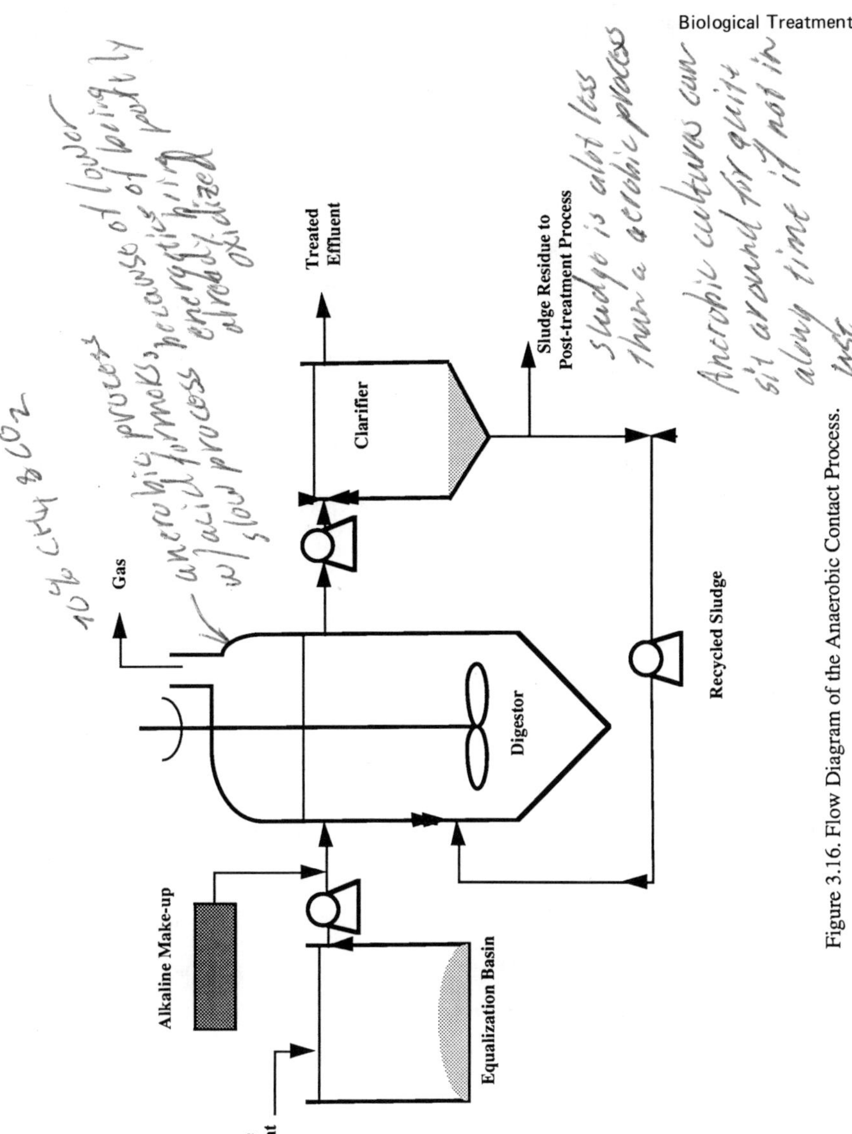

Figure 3.16. Flow Diagram of the Anaerobic Contact Process.

Table 3.9 Characteristics of Anaerobic Processes

Advantages	Disadvantages
• Low production of waste biological solids	• Little practical experience has been gained in full scale operations.
• Low nutrient requirements	• Relatively long periods are required to startup the process
• No aeration equipment required	• Anaerobic processes are limited to pretreatment applications; Additional treatment would be required to meet most water quality standards
• 90% conversion of biodegradable COD into methane, a useful by-product	• Process optimization for various types of wastes needs to be done
• Seasonal and intermittent operation possible; the sludge can be stored for long period of time	• Sensitivity to variable loads and possible toxicity problems
• High degree of waste stabilization (90%) possible at very high loadings	

Source: [10]

The process depends on the symbiotic relation of two classes of microorganisms: acid-forming bacteria and methane-forming bacteria. Facultative and anaerobic acid-forming bacteria first convert complex organic substrates in the wastes to short-chain organic acids (primarily acetic acid, propionic, and lactic acids), alcohols, carbon dioxide and H_2. Then strictly anaerobic methane-forming bacteria convert the volatile acids to methane gas, CO_2 and other trace gases. Fig. 3.17 shows the process of anaerobic degradation of organic compounds. Methane-forming bacteria are inherently slow growing, with doubling times measured in days [22]. They are also very sensitive to changes in the environment. In contrast, the acid-forming bacteria can function over a wide range of environmental conditions and have doubling times measured in hours [22]. When an anaerobic digestion is stressed by sudden changes in organic loads, temperature fluctuations, or an inhibitory material, the activity of methane-forming bacteria begins to lag behind that of acid-forming bacteria. The acids cannot be converted as rapidly as they form and pH drops.

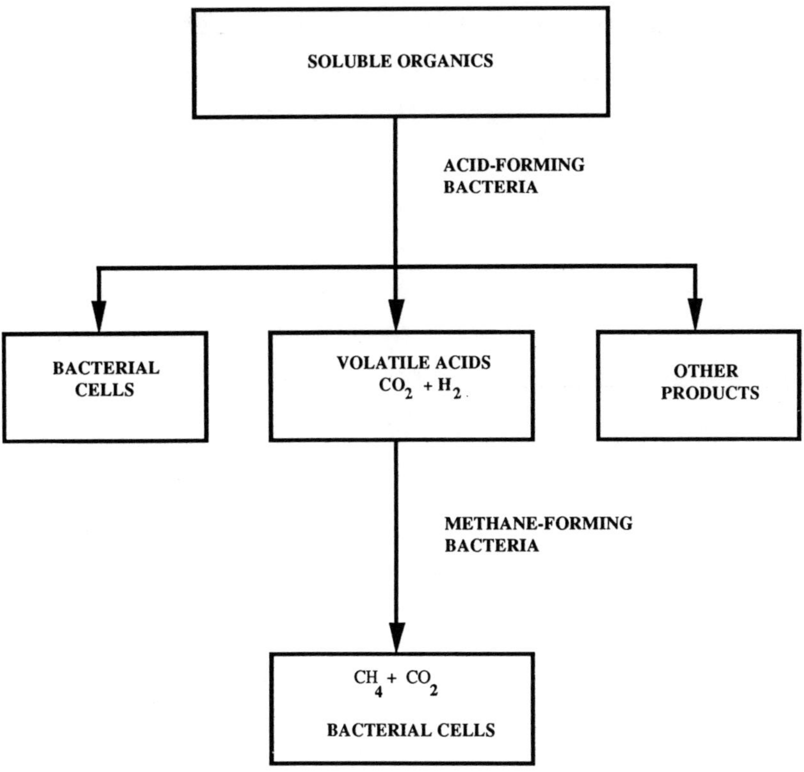

Fig. 3.17. Anaerobic Digestion of Organic Wastes.

The methanogens are further inhibited, and the process eventually fails. Therefore, the overall rates of anaerobic processes are controlled by methane-forming bacteria. When acid builds up, lime or bicarbonate may be added to control the pH. However, the best way is to stop the influent waste, and to allow the methane-forming bacteria to restore balance in the process.

The process is usually operated in a pH range of 6.8 to 7.5 and temperature range of 31 to 35°C [9]. However, some high-rate anaerobic contact digestors are operated at about 37°C to increase the rate of microbial growth. Since the methane-forming bacteria are recognized as the most sensitive microorganisms in anaerobic digestion, inhibition is indicated by the rate of methane gas production. Soluble heavy metals can cause the anaerobic digestor to fail. Inhibition of anaerobic digestion occurs at a soluble concentration of approximately 3 mg/l for Cr(+6), 2 mg/l for Ni, 1 mg/l for Zn, and 0.5 mg/l for Cu [22]. The light metal cations which come from industrial operations and the addition of alkaline material for pH control also play an important role in anaerobic digestion. They can be either stimulatory or toxic depending on their concentrations in solution. The stimulating and inhibitory concentrations of several light metal cations are shown in Table 3.10.

Table 3.10 Stimulating and Inhibitory Concentrations of Light Metal Cations

Cation	Concentration, mg/l		
	Stimulatory	Moderately inhibitory	Strongly inhibitory
Calcium	100 - 200	2500 - 4500	8000
Magnesium	75 - 150	1000 - 1500	3000
Potassium	200 - 400	2500 - 4500	12000
Sodium	100 - 200	3500 - 5500	8000

Source: [22]

The soluble heavy metals can be removed by the addition of sulfide compounds. Approximately 0.5 mg/l sulfide is needed to precipitate 1 mg/l of heavy metal. However, the soluble sulfides in the solution are toxic to the anaerobic digestion system if the concentration exceeds 200 mg/l [23]. Light metal cations may be detoxified by encouraging formation of carbonates and bicarbonates.

The free CN^- ion and chlorinated compounds have been found to be toxic in anaerobic systems at concentrations of 0.5 to 1.0 mg/l [24]. High concentrations of ammonia (over 3 g/l of total ammonia-nitrogen) can also cause problems and should be avoided [24]. The petrochemical industry is plagued by a number of compounds which, when present in wastestreams in concentrations above certain limits, pose a problem to anaerobic biological treatment. A list of hydrocarbons and their problem concentrations at which the methane gas production rate decreases by more than 50% is given in Table 3.11. These results show that inhibition is most pronounced in systems having high organic loadings (non-substrate limited).

For most industrial waste waters, nutrients such as nitrogen and phosphorus have to be added. The nitrogen requirement for anaerobic treatment is only a small fraction of that required by the aerobic process as indicated by the calculations done by Speece [21]. The phosphorus requirement is approximately 15% of the nitrogen requirement. Since the methanogens are unique in the anaerobic digestion process, they also need some unique trace nutrients. Studies have shown that trace amounts of iron, cobalt, nickel, sulfide, molybdenum, tungsten or selenium can stimulate the methane gas production rate [21].

Generally if an industrial wastewater is treatable aerobically, it will be treatable anaerobically, although there are exceptions. For example, an anaerobic digestion system is not capable of significant decomposition of hydrocarbons with long chains or benzene rings. These organics are degraded by oxygenase enzymes present in aerobic microorganisms, but not in anaerobes. Table 3.12 is a list of organics that have been demonstrated to be anaerobically biodegradable and that are potential components of industrial wastewaters.

Table 3.11 Problem Concentrations of Some Hydrocarbons in Anaerobic Digestion

Chemical	Problem Concentration (mg/l) Substrate Limiting	Non-Substrate Limiting
n-Butanol	--	>1000
sec-Butanol	--	>1000
t-Butanol	>1000	>1000
Allyl alcohol	--	>1000
2-Ethyl-1-hexanol	500-1000	--
Formaldehyde	--	50-100
Crotonaldehyde	~200	50-100
Acrolein	--	20-50
Acetone	--	>1000
Methyl isobutyl ketone	>1000	100-300
Isophorone	>1000	--
Diethylamine	--	300-1000
Ethylene diamine	--	100-300
Acrylonitrile	150-500	100
2-Methyl-5-ethylpyridine	>1000	100
N,N-dimethylaniline	>1000	--
Phenol	>1000	300-1000
Ethyl benzene	>1000	--
Sodium benzoate	--	>300
Ethylene dichloride	150-500	--
Ethyl acrylate	600-1000	300-600
Sodium acrylate	--	>500
Dodecane	>1000	--
Dextrose	>1000	>1000
Ethyl acetate	>1000	--
Ethylene glycol	>1000	>900
Diethylene glycol	--	>1000
Tetralin	>1000	--
Kerosene	--	>500
Cobalt chloride	--	>1000

Source: [9]

Table 3.12 Organics Amenable to Anaerobic Biotechnology

Acetaldehyde	Acetic anhydride	Acetone
Acrylic acid	Adipic acid	Aniline
1-amino-2-propanol	4-amino butyric acid	Benzoic acid
Butanol	Butyraldehyde	Butylene glycerol
Catechol	Cresol	Crotonaldehyde
Crotonic acid	Diacetone gulusonic acid	Dimethoxy benzoic acid
Ethanol	Ethyl acetate	Ethyl acrylate
Ferulic acid	Formaldehyde	Formic acid
Fumaric acid	Glutamic acid	Glutaric acid
Glycerol	Hexanoic acid	Hydroquinone
Isopropanol	Lactic acid	Maleic acid
Methanol	Methyl acetate	Methyl acrylate
Methyl ethyl ketone	Methyl formate	Nitrobenzene
Pentaerythritol	Pentanol	Phenol
Phthalic acid	Propanal	Propanol
Isopropyl alcohol	Propionate	Propylene glycol
Protocatechuic acid	Resorcinol	sec-butanol
sec-butylamine	Sorbic acid	Syringaldehyde
Syringic acid	Succinic acid	Tert-butanol
Vanillic acid	Vinyl acetate	

Source: [21]

3.2.3.b. Cost Estimation

The literature construction cost data for the anaerobic digestor are calculated based upon the following design requirements:

a heat exchanger heats incoming wastestream to digestor temperature,

an electrical stirrer is used for mixing,

20 ft submergence for release of gas.

The results within the same influent flow rate range as that used in the aerobic processes are shown in Fig. 3.18 [15]. The annual O&M costs for the digestor unit are shown in Fig. 3.19 [15].

Because of the ability to handle high solids content wastestreams, the anaerobic treatment process usually does not require a primary clarifier in its upstream pretreatment process. Only the flow equalization basin is used in the pretreatment process. The secondary clarifier in the effluent stream, however, is still needed to separate the solids from the treated water. Estimates of

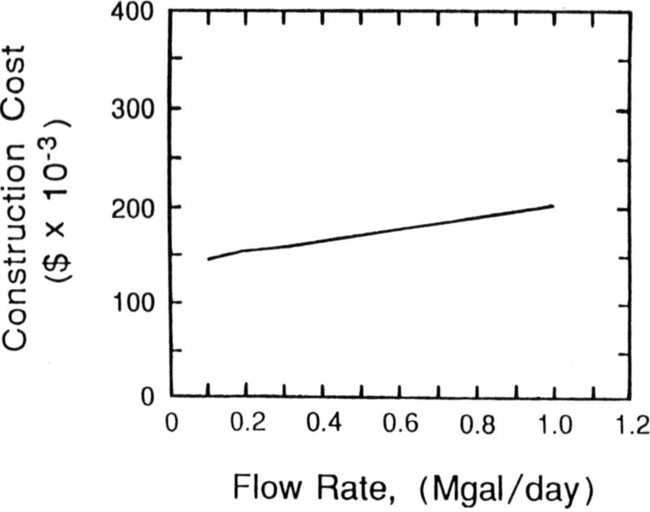

Figure 3.18. Construction Cost for Anaerobic Digestor.

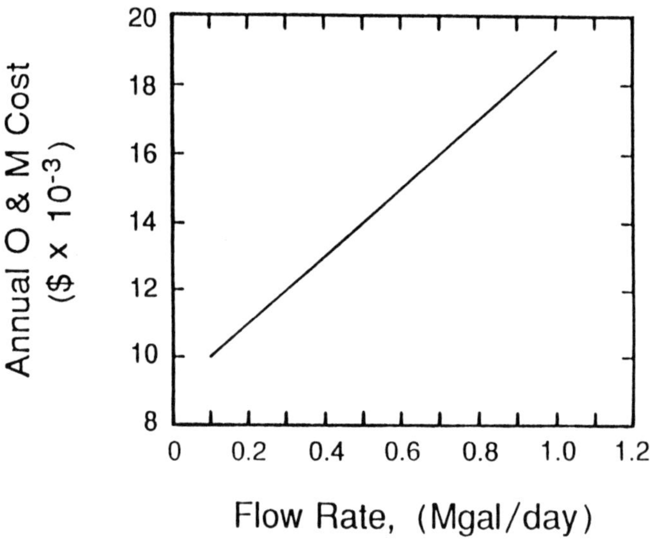

Figure 3.19. Annual O&M Cost for Anaerobic Digestor.

construction and O&M costs for the flow equalization basin and the secondary clarifier, described in the section on the activated sludge process, were used with cost estimates for the anaerobic digestor, to obtain estimates of the total construction cost and annual O&M cost of the process. The capital costs and the annual capital costs are calculated following the same procedures and conditions used in the aerobic processes. The construction costs for influent flow rates between 0.1 and 1 Mgal/day are shown in Fig. 3.20. The annual O&M costs and operating costs for the same flow rate range are shown in Figs. 3.21 and 3.22 respectively.

3.3. ASSESSING THE TREATABILITY OF CALIFORNIA WASTES BY BIOLOGICAL TREATMENT PROCESSES

A database program was written using Fox BASE (Fox Software) to determine which of the wastestreams found in the 1985 Biennial Generator Report (BGR) database [25] (California) are suitable for biological treatment. The costs of four treatment processes for each treatable wastestream were also estimated in the program. The algorithm of the program is explained in a flow chart shown in Fig. 3.23. Details of the program are discussed in the following sections.

3.3.1. Results on Treatability Using Biological Treatment

Because only aqueous wastestreams with low solid content (<1%) can be treated by biological treatment, the program first examined the physical state of a wastestream found in the database. Only those wastestreams denoted as liquid or sludge were considered. Because the database lacks quantitative data on solids content, streams were required to contain at least 90% water as a means of ensuring that both solids content and organic content were low.

For streams which passed the above tests, the name of each component in the wastestream was converted to its unambiguous counterpart using the translator database. The "converted" name was then used to check the chemical characteristics of the component in the property database. The concentrations of halogenated organics, organics, aromatics, and cyanides were summed individually and the results were compared with the following treatability limits:

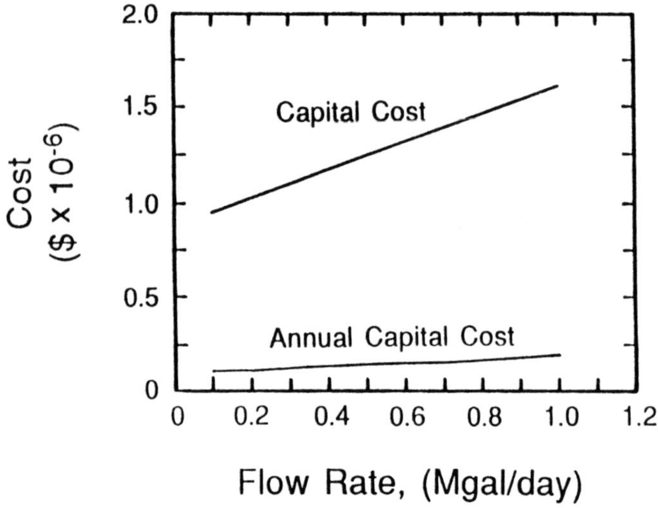

Figure 3.20. Capital and Annual Capital Costs for Anaerobic Contact Process.

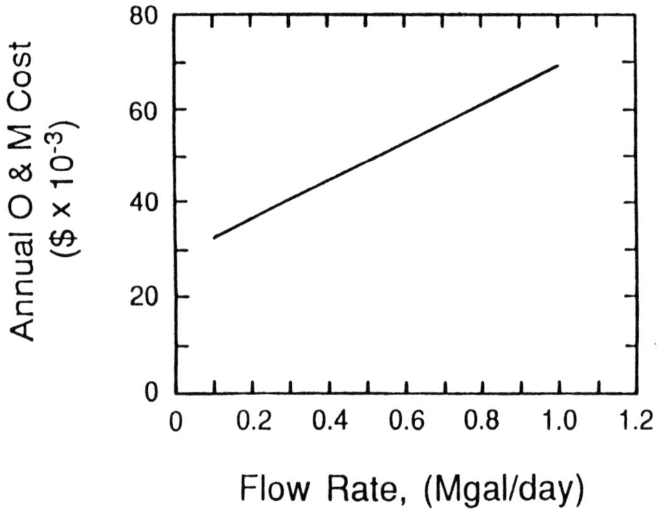

Figure 3.21. Annual O&M Cost for Anaerobic Contact Process.

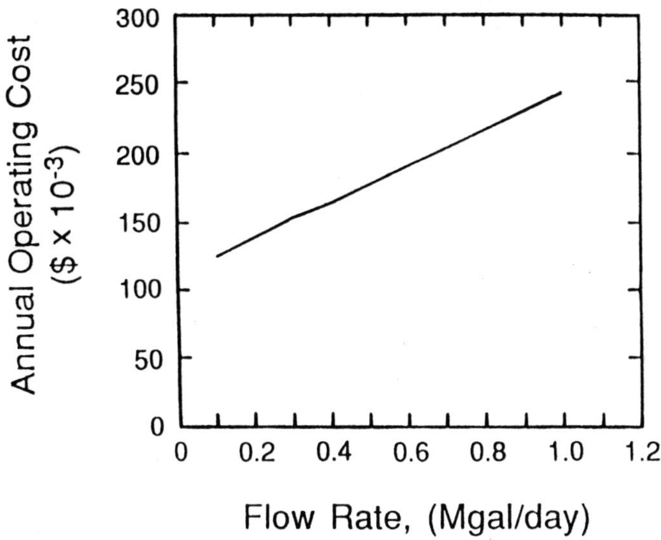

Figure 3.22. Annual Operating Cost for Anaerobic Contact Process.

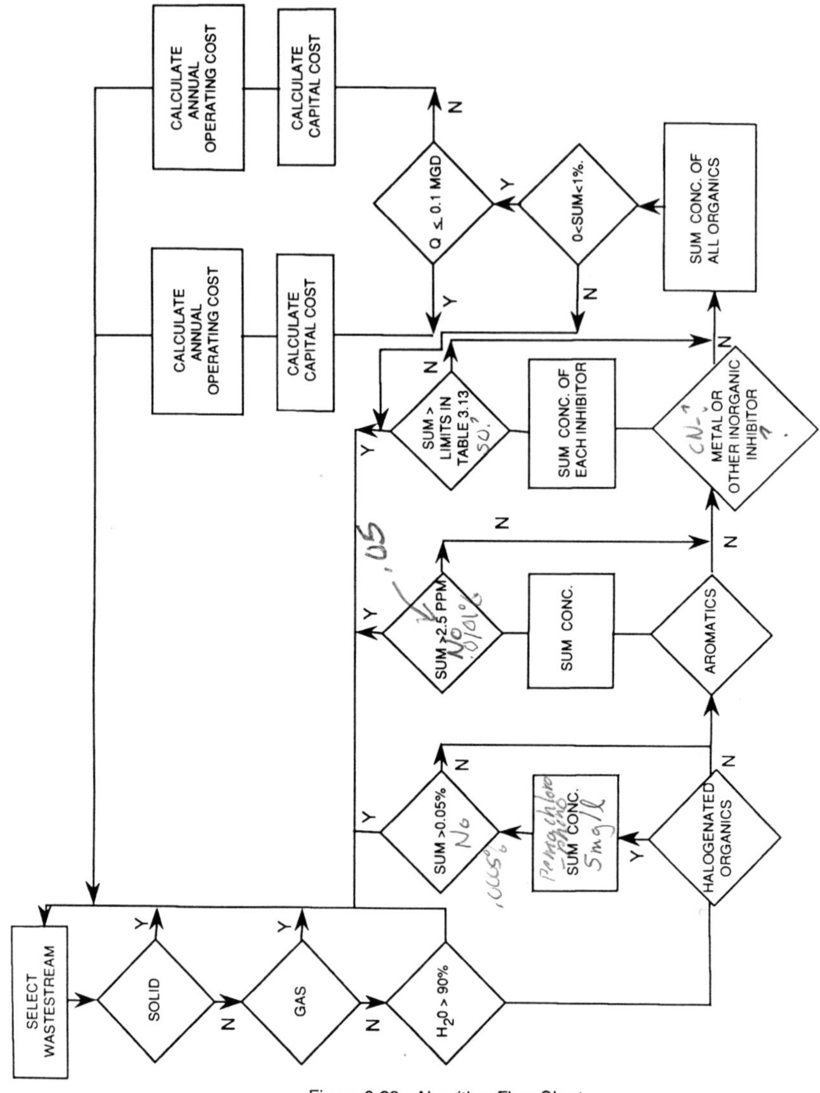

Figure 3.23. Algorithm Flow Chart

halogenated organics	0.05%
organics	1%
aromatics	0.05%
cyanide compounds	2.5 ppm

Any wastestream that had a concentration of any of the above components higher than the limit was deemed unsuitable for biological treatment. Wastestreams with zero organic concentration were also unsuitable.

The program then determined whether the wastestream contained any of the metals listed in Table 3.13 by searching for the metals' chemical symbols in the field name Chemical Formula in the property database. When the component was found to contain one or more of those metal inhibitors, the individual metal concentration of this component in the wastestream was calculated. After checking out all the components, the program summed the concentration of each metal inhibitor in the wastestream. The results were compared with the limits shown in Table 3.13. If any one those metal concentrations exceeded the limit, the wastestream was considered unsuitable for biological treatment.

Those wastestreams that passed the above checks were considered suitable for biological treatment. The information on those streams was stored in a separate database file that contains information on each treatable stream found in the BGR database.

Due to the complexity of biological treatment processes, this program did not attempt to determine which biological treatment process is most suitable for a given wastestream.

3.3.2. Cost Estimation

After the wastestreams suitable for biological treatment were identified, the capital and annual operating costs were calculated from the literature data discussed in Sections 3.2.1.b, 3.2.2.b, and 3.2.3.b. Because the cost data obtained from the literature are functions of influent flow rate, the annual discharge volume (AVG_VOL) reported in BGR database had to be converted to the volumetric flow rate. It was assumed that all facilities operate for 250 days each year and

Table 3.13 Toxicity-Threshold Concentrations for Various Components in Biological Treatment

Component	Concentration (mg/l)
Silver	0.03
Vanadium	10.0
Zinc	2.0
Nickel	1.75
Chromium	10.0
Lead	0.1
Iron(Ferric)	15.0
Copper	5.5
Cadmium	1.0
Arsenic	0.1
Boron	50
Manganese	10
Mercury	2.5
Calcium	2500
Magnesium	1000
Potassium	2500
Sodium	3500

that the density of all wastestreams is the density of water (0.0042 ton/gal). The conversion equation can thus be written as follows:.

$$Q \text{ (Mgal/day)} = \frac{\text{AVG\_VOL (tons/yr)}}{(250 \text{ day/yr})\ (0.0042 \text{ ton/gal}) \times 10^6} - 9.6 \times 10^{-7} \times \text{AVG\_VOL} \qquad (3.2)$$

The smallest flow rate for which literature cost data can be found for all four processes is 0.1 Mgal/day. Therefore, the capital cost was assumed to be the same as that for 0.1 Mgal/day for all wastestreams with flow rate less than 0.1 Mgal/day. For the wastestreams having flow rate higher than 0.1 Mgal/day, capital cost was calculated using the linear correlation equations obtained from Figs. 3.4, 3.12, and 3.20. The correlation equations and the minimum capital costs for each of the biological processes are listed below.

For Q $\leq$ 0.1 Mgal/day

Capital cost = 0.91 million (activated sludge process)

= 0.76 million (trickling filter process)

= 1.136 million (RBC process)

= 0.922 million (anaerobic contact process)

For Q > 0.1 Mgal/day

Capital cost = 0.806 + 1.0 * Q (activated sludge process) (3.3)

= 0.643 + 1.185 * Q (trickling filter process) (3.4)

= 0.976 + 1.598 * Q (RBC process) (3.5)

= 0.853 + 0.697 * Q (anaerobic contact process) (3.6)

The annual operating cost is the sum of the annual capital cost and annual O&M cost. The correlation equations obtained from the data shown in Figs. 3.6, 3.15 and 3.22 represent operating costs for Q>0.1 Mgal/day. These equations are presented as follows:

Annual operating cost = 0.112 + 0.155 * Q (activated sludge process) (3.7)

= 0.090 + 0.164 * Q (trickling filter process) (3.8)

= 0.136 + 0.224 * Q (RBC process) (3.9)

= 0.121 + 0.119 * Q (anaerobic contact process) (3.10)

For the case of Q$\leq$0.1 Mgal/day, the annual capital cost was first obtained from the above minimum capital cost data using Eq. (3.1), then added to the correlation equations of O&M cost data shown in Figs. 3.4, 3.14 and 3.21 to obtain the correlation equations for the annual operating costs. The resulting correlation equations are shown as follows:

Annual operating cost = 0.123 + 0.045 * Q (activated sludge process) (3.11)

= 0.103 + 0.033 * Q (trickling filter process) (3.12)

= 0.153 + 0.048 * Q (RBC process) (3.13)

$$= 0.129 + 0.042 * Q \quad \text{(anaerobic contact process)} \tag{3.14}$$

3.3.3. Results of Analysis of the Suitability of Biological Treatment

The results of the program are summarized in Table 3.14 which contains data on wastestreams treated by biological treatment processes.

Only 129 wastestreams out of total 9938 wastestreams in the BGR database are treatable by biological treatment processes. About 38% of the treatable wastestreams are generated by the chemical and petroleum refining industries (SIC codes 28 and 29), and another 10% by the transportation equipment industry (SIC code 37). These results are reasonable because the aqueous wastes generated by these industries often contain mainly organic compounds at levels that can be effectively handled by biological treatment. The rest of the treatable wastestreams are generated by a wide range of different industries, such as communications (SIC Code 48), metal-finishing industries (SIC codes 33, 34, 35 and 36) and printing and publishing industries (SIC code 27).

Because the flow rates of all treatable wastestreams are very low (much less than 0.1 MGD), the treatment costs were estimated at the fixed minimum values. These results are listed in Table 3.15. The capital costs range from the highest of $1.136 million for the RBC process to the lowest of $0.76 million for the trickling filter process. The highest and lowest annual operating costs are also found to be the RBC process ($0.153 million) and trickling filter process ($0.103 million) respectively.

Table 3.14 List of Wastestreams That Can Be Treated By Biological Treatment

SIC	Volume (Tons)	Water (%)	Organic (%)	Generating Process
	50	99.2	0.63	Paper corrugation
10	87.5	99	1	Flushing of gas storage tank prior to removal
10	10.42	99	1	Flushing inactive gasoline storage tank prior to destruction
13	1429	99.5	0.5	Oil well drilling
13	131	99.7	0.3	Gasoline/gas separating facility
13	142	99.5	0.5	Oil production operations
24	71.25	99	1	Removal of discarded fuel oil tank
27	3.75	95	1	Bulk storage tank cleaning
27	110.42	92.2	0.3	Cleaning press part and ink pumps
27	5	99.93	0.06	Waste water from agricultural chemical production
27	3.44	93.01	0.51	Electrostatic precipitation
28	27	99	1	Rain water run off into containment area
2810	37	98.06	0.25	Plating, nickel, chrome
2820	10.42	99	1	Site cleanup material
2820	25	98	1	Solution coating of synthetic substrates
2820	162.2	99.8	0.2	Under water cleanup
2820	15	97	1	Synthetic resin impregnation of synthetic substrates
2820	120	90.78	0.48	Cleanup water from water based primer paint line
2840	26.6	98	0.5	Industrial sewer sludge
2851	83.33	98.5	1	Produced when washing resin kettle between batches
2851	14	98.48	0.42	Manufacturing of dispersions
2860	73	98	1	Tank bottom sediment
2860	2	99	1	Tank bottom sediment
2870	40	99	1	Wash out surfactant mixing equipment
2870	16725	99	1	Waste water from agricultural chemical production
2870	7.25	99.96	0.04	Site cleanup material
2879	50	99.97	0.03	Rinsate from agricultural spraying equipment
2890	32.92	99.9	0.1	Drum residual/blending tank rinsing
2890	398	98.9	0.1	Pharmaceutical process wash water
2890	12007	98.9	0.1	Process wash down water and lab drains
2891	12.5	99.95	0.05	Adhesive mixer wash
29	40.83	96	1	Chemical cleaning
29	109.76	96.7	0.3	Chemical cleaning
29	4.58	98	1	Chemical cleaning
29	129.17	99.93	0.07	Crude & fuel oil dehydration
29	50	95	1	Tank cleaning
29	8	99.5	0.5	Tank cleaning
29	48.5	99.5	0.5	Site cleanup/tank replacement
29	10.3	99.5	0.5	Tank cleaning
29	6.4	99.5	0.5	Tank cleaning
29	3.2	99.5	0.5	Tank replacement/site cleanup

(continued)

Table 3.14 (continued)

29	5.7	99.5	0.5	Tank cleaning
29	40	99.5	0.5	Site cleanup/tank replacement
29	3	99.5	0.5	Site cleanup/tank replacement
29	77.8	99.5	0.5	Site cleanup/tank replacement
29	11.8	99.5	0.5	Site cleanup/tank replacement
2911	1000	93.39	0.94	Effluent treatment plant
2911	86.19	95	1	Equipment cleanup material
2911	438.38	95	1	Equipment cleanup material
2911	15.42	95	1	Equipment cleanup material
2911	29.13	95	1	Equipment cleanup material
2911	36.2	95	1	Equipment cleanup material
2911	16.8	95	1	Equipment cleanup material
2911	13.5	95	1	Equipment cleanup material
2911	20.99	95	1	Equipment cleanup material
2911	136.1	95	1	Equipment cleanup material
2911	51.58	95	1	Equipment cleanup material
2911	362	95.06	0.19	Cleanup from wastewater treatment facilities
33	75	98.85	0.5	Caustic cleaning of steel strip
3470	1	99.95	0.02	Oil quench bath residue
3470	229.17	97.75	1	Oil, flux, floatable material skimmed from neutralizer
35	54.38	99.97	0.03	34,35 monitoring well sampling and testing
35	60.42	99.97	0.03	34,35 monitoring well pumping, testing
36	16.43	98	1	Chemical storage tank cleaning
36	3.21	n .48	0.5	Decommissioning R&D electroplating lab
36	2.57	98.62	0.35	Spent plating bath solution
36	7.1	99.15	0.21	Manufacture of printed circuit boards
3674	~0	99.99	0.01	Spill
3679	1.83	93	1	Printed circuit manufacturing process
3679	0.92	95	1	Printed circuit manufacturing process
3679	1	99.95	0.05	Waste water
3710	1761	99.94	0.02	Waste water treatment facility for coating operations
3720	23.33	95	1	90 day clarifier clean according to SQAMD requirements
3720	978	99	1	Rinse water from overflowing rinse tanks
3720	10	99	1	Lube oil-contaminated water from metal forming machines
3720	280	99	1	Spill cleanup
3720	20.83	99.5	0.5	Spill cleanup
3721	1.25	99.96	0.02	Clarifier sludge
3728	8.33	99	1	Mfg. process
3728	0.23	99.95	0.02	Unknown 55 gal drum C
3730	1650	99	1	Removing oil/water ballast from ocean vessels
3730	1850	99.48	0.5	Cleaning of barges, tugs and oil spill containment boom
3760	148.49	99.97	0.03	Machining of rocket engine parts
3760	250	99.93	0.06	Cleanup clarifiers serving water wash paint booths, steam cleaner
3760	241.67	99.7	0.3	Waste coolant from machining of parts
39	4.2	99	1	Underground fuel tank
39	12	95.15	0.1	Deburring of aluminum castings
40	82.08	98.88	0.7	Railcar parts cleaning
42	257.08	99.5	0.5	Waste water from wash rack
42	12.3	98	1	Steam cleaner residue

Table 3.14 (continued)

48	11.67	99.78	0.18	Waste water removed from a vehicle stream cleaning sump
48	389	99	1	Leaking underground storage tank
48	6.25	99	1	Tank wash out before removal
48	10.42	97	1	Wash out of tank before removal
48	13	99	1	Pumpings from manholes
48	25	99	1	Pumpings from manholes
48	3	99	1	Pumpings from manholes
48	24	99	1	Pumpings from manholes
48	20.9	99.9	0.1	Pumpings from manholes
48	13	99	1	Pumpings from manholes
48	6	99	1	Pumpings from manholes
48	7	99	1	Pumpings from manholes
48	0.1	99	1	Pumpings from manholes
49	3.2	99.99	0.01	Drain cooling system
49	92.4	99	0.5	Flushing contaminated tank
49	33.2	98.51	0.5	Tank flushing
49	32.7	99	0.5	Tank flushing
49	30.2	99	0.5	Tank flushing
49	10.4	~ .00	0.5	Tank flushing
4911	12.5	99.5	0.5	Steam cleaning waste from cleaning a gasoline truck
51	20	99.99	0.01	Mechanical zinc plating
5541	181.14	99	1	Service station
5541	55	99	1	Service station
73	6.67	99	1	Residue from cleanup of empty drums
75	10	95	1	Steam cleaning residue - mud & water
95	8.33	99.8	0.1	Rinse water for cleanup in rodenticide formulation process
97	41.67	99	1	Slop collection tank
97	5	90.45	0.5	Cleanup of fuel spill from refueling equipment
97	2841.67	99.9	0.1	Chemical toilet cleanout
97	0.8	99	1	Spill cleanup
99	45	99	1	Natural seepage
99	0.46	99.97	0.03	Mixed solvent/cleanup material
99	528	99.96	0.04	Fire and water damage/site cleanup
99	10.5	95	1	Washing of commercial linens & uniforms with detergents/soaps
99	120	99	1	Yard dump
99	10.8	99.9	0.1	Rainfall into contaminated excavation
99	2.62	99	1	Removal of 1,000 gal gasoline storage tank
99	5	96	0.5	Equipment wash area

Table 3.15 Capital and Annual Operating Cost Data for the Biological Treatment Processes

	Activated Sludge process	Trickling Filter process	RBC process	Anaerobic Contact process
Capital Cost (million $)	0.910	0.760	1.136	0.922
Annual Operating Cost (million $)	0.123	0.103	0.153	0.129

3.4. REFERENCES

1. "Biodegradation Techniques for Industrial Organic Wastes," edited by D.J. De Renzo, Noyes Data Corp., Park Ridge, New Jersey, (1980).
2. Winkler, M., "Biological Treatment of Waste-Water," Ellis Horwood Ltd., England, (1981).
3. "Alternative Technology for Recycling and Treatment of Hazardous Wastes," 3rd biennial report, edited by J. Potter, Dept. of Health Services, State of California, (1986).
4. Canney, P.J., C.J. Mordorski, R.M. Rollins and C.L. Berndt, "PACT Wastewater Treatment for Toxic Waste Cleanup," 39th Purdue Conference, 413 (1984).
5. Sutton, P.M., "Innovative Engineered Systems for Biological Treatment of Contaminated Surface and Groundwater," 7th National Superfund Conference Proceedings., Washington D. C. (1986).
6. Hemming, M.L., "General Biological Aspects of Waste-Water Treatment Including the Deep-Shaft Process," Wat. Pollut. Control, 312 (1979).
7. Bridie, A.L., Wolff, C.J.M., and Winter, M., "BOD and COD of Some Petrochemicals," Water Research (Brit.), 13, 627 (1979).
8. Junkins, Randy, K. Deeny, and T. Eckhoff, "The Activated Sludge Process: Fundamentals of Operation," Ann Arbor Science, Ann Arbor, Michigan, (1983).
9. Arthur D. Little, Inc., "Physical, Chemical, and Biological Treatment Techniques for Industrial Wastes," report to EPA, (1977).
10. Kobayashi, H. and B.E. Rittmann, "Microbial Removal of Hazardous Organic Compounds," Environ. Sci. Technol., 16, 3, 170A (1982).
11. Patterson, J.W. and P.S. Kodukala, "Biodegradation of Hazardous Organic Pollutants," Chem. Engng. Progress, 48, April (1981).

12. Metcalf & Eddy, Inc., "Technologies Applicable to Hazardous Waste," Report to EPA, (1985).

13. Ganczarczyk, J.J., "Activated Sludge Process, Theory and Practice," Marcel Dekker, Inc., New York (1983).

14. Dugan, P.R., "Biochemical Ecology of Water Pollution," Plenum Press, New York (1972).

15. "Handbook of Wastewater Treatment Processes," edited by A.S. Vernick and E.C. Walker, Marcel Dekker, Inc., New York (1981).

16. "Economic Indicators," Chemical Engineering, p. 9, McGraw Hill, New York, February 15, 1988.

17. Benjes, H.H., "Handbook of Biological Wastewater Treatment," Garland STPM Press, New York (1983).

18. Eckenfelder, Jr., W.W., "Water Quality Engineering for Practicing Engineers," Barnes & Noble Inc., New York (1970).

19. Borchardt, J.A., "Biological Waste Treatment Using Rotating Discs," in Biological Waste Treatment, ed. Canale, R.P., Wiley-Interscience, New York (1971).

20. Obayashi, A.W., H.D. Stensel and E. Kominek, "Anaerobic Treatment of High-Strength Wastes," Chem. Engng. Progress, 68, April (1981).

21. Speece, R.E., "Anaerobic Biotechnology for Industrial Waste Water Treatment," Environ. Sci. Technol., 17, 416A (1983).

22. "Process Design Manual — Sludge Treatment and Disposal," Environmental Protection Agency (1979).

23. Lawrence, A.W. and P.L. McCarty, "The Role of Sulfide in Preventing Heavy Metal Toxicity in Anaerobic Treatment," Journal WPCF, 37, 392 (1965).

24. Lettinga, G., Biotech. and Bioeng., 22, 699 (1980).

25. R.L. Bell, A.P. Jackman and R.L. Powell, a report submitted to the California Dept. of Health Services by Dept. of Chemical Engr., U. of California, Davis, 1987.

26. Stover, E.L. and D.F. Kincannon, "Biological Treatability of Specific Organic Compounds Found in Chemical Industry Wastewaters," Journal WPCF, 55, 1, 97 (1983).

27. Wheatley, A.D., "Biotechnology and Effluent Treatment," in Biotechnology and Genetic Engineering Reviews, p. 261, Vol. 1, ed. G. E. Russell, Intercept Ltd. (1984).

4. Wet Air Oxidation

4.1. INTRODUCTION

Wet Air Oxidation (WAO) has become a popular process for treating aqueous wastes containing both organic and inorganic toxics. It is used to economically treat wastestreams too dilute to incinerate and too toxic to biotreat. The function of WAO process is to alter the chemical structure by low temperature oxidation of the waste so that toxic compounds become nontoxic. It is generally considered to be a pretreatment process. The WAO process has been employed in ethylene, olefin, acrylonitrile, steel, petrochemical, and herbicide production plants. It is excellent for night soil, cyanide and spent caustic liquors reduction [1].

Zimpro Inc. developed the Wet Air Oxidation process, and is the supplier of the majority of published information concerning the process and its performance. As of 1982, there were approximately 200 installations of the WAO process [2]. Several companies which installed the WAO process have published their results. The Casmalia Class I hazardous waste disposal site in California uses WAO to treat wastes which contain cyanides, sulfides, phenols, solvent still bottoms, and organic waste waters. Bofors-Nobel Inc., Michigan, applies the process to treat wastes from chemical manufacturing plants. Dominion Foundries and Steel Co., Ontario, Canada, detoxifies coke oven gas. Dominion's process produces ammonium sulfate which is saleable as fertilizer. Other corporations which have installed the WAO process include: Northern Petrochemical Company, Illinois; Uniroyal Chemicals Ltd., Ontario, Canada; and Mitsubishi Petrochemical Co., Japan.

As the number of WAO units has increased, the understanding of the chemistry of the process has improved. The addition of catalysts has been shown to improve removal efficiencies and/or lower required operating temperatures. There has been much interest in finding construction materials which can better withstand the harsh internal operating environment. And the improved understanding of how process parameters may be adjusted to control the reaction has resulted in more stable operation of WAO units. As process improvements have been made, the process has become an increasingly attractive technology for hazardous waste treatment.

4.2. DESCRIPTION OF THE WAO PROCESS

Wet air oxidation is a liquid phase process, although solids and/or gases may be present in the wastestream. The basic flow diagram for the process is shown in Figure 4.1. The waste enters the system at ambient temperature through a high pressure pump. Units presently operating are designed for flow rates ranging from 6 gal/min to 3,000 gal/min [1]. The waste is then sent through two heat exchangers, usually on the tube side, in order to raise the temperature of the waste. The first, usually a single-pass countercurrent exchanger, recovers heat from the process effluent. This preheating is necessary in order to assure that the waste will reach the temperature needed for complete conversion of the waste while in the reactor. A second exchanger, usually oil fired, is required for startup and is sometimes used for auxiliary heating during normal operation. If the required chemical oxygen demand (COD) reduction is 15,000 to 20,000 mg/L or higher and a heat exchanger is employed, then no auxiliary heating source is needed once the process reaches steady state. If the required COD reduction is less than 15,000 mg/L, auxiliary energy is supplied continuously by the hot oil heater. The maximum requirement for auxiliary energy is between 500 and 600 Btu/gal [3]. Compressed air or oxygen is injected into the heated waste, at a concentration which depends on the concentration of the oxidizable waste components. The greater the concentrations of the waste components, the greater the amount of oxygen needed. The WAO process uses a continuous flow tubular reactor. Both the waste and air enter the reactor simultaneously at the inlet to the reactor. There is little published information on sizes of full scale, commercial reactors. The residence time used in the laboratory tests is typically 60 min. By adjusting the temperature and pressure in the reactor, it is frequently possible to obtain 99% conversion of toxic chemicals in the waste. More information on conversions is given below.

The effluent exits from the reactor and is sent through the shell-side of the heat exchanger which preheats the feed. This utilizes the heat evolved from the oxidation reaction to heat the

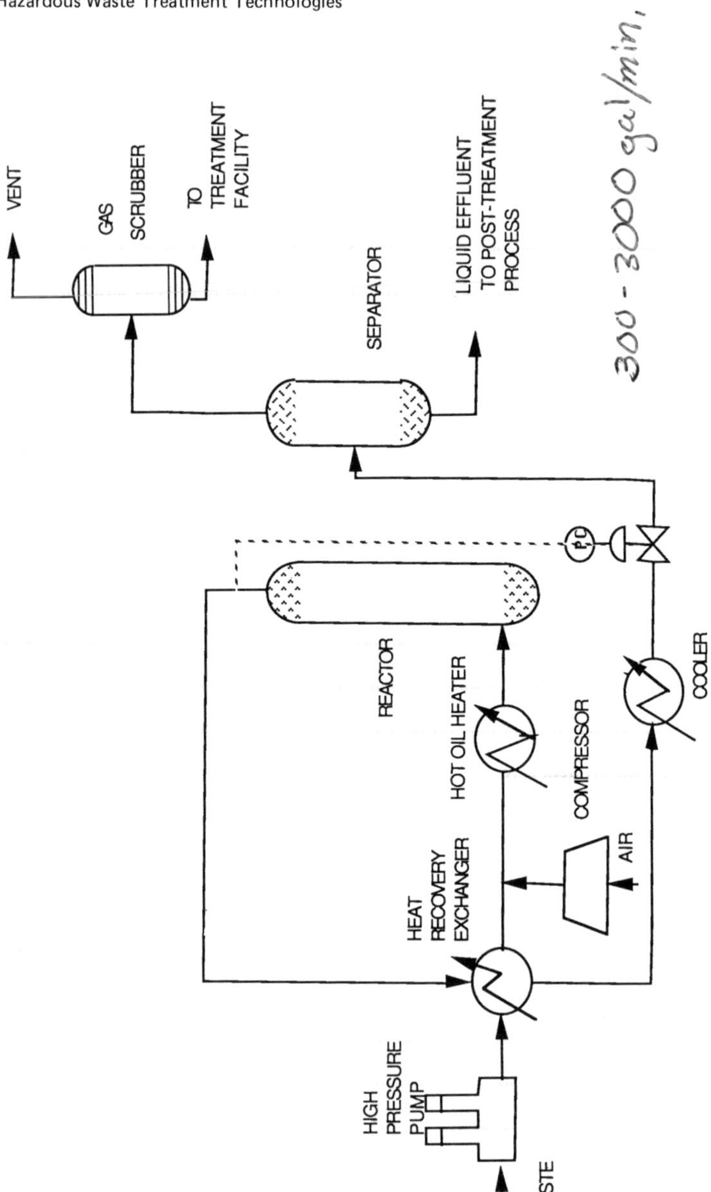

Figure 4.1. Schematic of Wet Air Oxidation Process

incoming waste. If the feed preheater is not employed, a maximum of 4,000 Btu/gal [3] of auxiliary energy from the hot oil heater is needed to heat the incoming waste. The waste is then cooled further through a cooler followed by a separator.

Depending on the waste, the off-gas may have to be scrubbed to remove any low molecular weight hydrocarbons present such as acetaldehyde. Treatment of wastes from enhanced oil recovery and ethylene manufacturing has been found not to require off-gas scrubbing. However, most existing installations do include off-gas scrubbing. Alternatives to scrubbing are carbon adsorption and fume incineration [3].

In the liquid stream, low molecular weight compounds will also be present which need to be further treated. Because 100% conversion of the toxic waste to carbon dioxide and water may not occur when using WAO, the process is sometimes followed by another treatment. Usually it is a biodegradation process. Solid products may also be present in the liquid. These are separated from the stream and disposed of using a fixation or solidification process.

In some cases, an ionic catalyst may be added to improve conversion efficiency. If an ionic catalyst is added to the reactor, a catalyst regenerator must be added to the process. The catalyst regenerator removes the ions from the effluent stream and returns them to the reactor. Removal of the catalyst from the effluent is required to prevent contamination as well as to reduce the operating cost. Also, a controller system may be installed for automating the process.

4.2.1. Treatability of Compounds

There is much published information on the various classes of compounds which may be efficiently treated by WAO. These results are summarized in Table 4.1. Ideally, organic compounds are oxidized stoichiometrically with carbon being converted to carbon dioxide, hydrogen to water, halogens to acid halides, sulfur to inorganic sulfate, phosphorus to phosphate, and nitrogen to ammonia, nitrate, or elemental nitrogen. Table 4.2 lists those compounds which have been reported in the literature to be oxidized using the WAO process. The data include WAO

Table 4.1. Toxic Contaminants Treatable by Wet Air Oxidation

- Inorganic and organic cyanide compounds
  (Toluene Diisocyanate, Cyanide)

- Aliphatic and chlorinated aliphatic compounds
  (Carbon Tetrachloride, Acetaldehyde)

- Aromatic hydrocarbons
  (Toluene, Acenaphthene, and Pyrene)

- Aromatic and halogenated aromatic compounds containing non-halogen functional groups
  (Pentachlorophenol, 2,4,6-Trichloroaniline)

- Heterocyclic compounds containing O, N, or S atoms provide a point of attack for oxidation
  (Mercaptans, Benzidine)

Source: [13]

performance on pure components, synthetic and actual wastes that were oxidized in bench or full scale reactor units.

The influent concentration of the treated components listed in Table 4.2 does not necessarily represent the upper limits for WAO treatment; it merely means that at such inlet concentrations and operating conditions, the removal efficiencies have been determined and are shown in Table 4.2. For higher influent concentrations than those found in Table 4.2, WAO may still be applicable, but the resulting removal efficiency may be lower if the operating conditions remain the same. In such cases, in order to achieve the same efficiency as listed in Table 4.2, the operating parameters will need to be re-adjusted (e.g., increasing the residence time, temperature, pressure, etc.). Thus, there is no definite influent concentration limit for the wet air oxidation process. However, for the purpose of assessing the feasibility of the WAO process in this study, we have chosen to group wastes according to influent concentration of oxidizable wastes as follows: 0-1% (grade=A, high

Table 4.2 Waste Components Treatable with WAO

COMPONENT	INFLUENT CONC. (MG/L)	EFFLUENT CONC. (MG/L)	FLOW RATE (GPM)	PERCENT REMOVAL	CATALYST REQUIRED	TEMP (C°)	PRESSURE (PSIG)	TIME (MIN)
1,1,1-trichloroethane	4400.0	44.0	--	99.00	yes	320	--	--
1,1-dichloroethylene	210.0	2.1	--	99.00	yes	320	--	--
1,1-dimethylhydrazine	9250.0	0.7	--	99.99	no	271	--	117
1,2-dichloro-3-nitrobenzene	--	--	--	80.60	no	285	--	120
1,2-dichlorobenzene	590.0	150.0	--	74.60	no	320	--	120
1,2-dichlorobenzene	2213.0	29.0	0.04	98.70	no	314	1934	128
1,2-dichlorobenzene	6530.0	2017.0	--	69.10	no	320	--	60
1,2-dichloroethane	6280.0	13.0	--	99.80	no	275	--	60
1,2-diphenylhydrazine	5000.0	6.0	--	99.88	no	275	--	60
1,2-diphenylhydrazine	5000.0	6.0	--	99.98	no	275	--	60
1,3-butadiene	735.0	7.4	--	99.00	yes	320	--	--
1,4-dioxane	20000.0	200.0	--	99.00	yes	320	--	--
1-chloronaphthalene	5970.0	5.0	--	99.92	yes	275	--	60
1-chloronaphthalene	1524.0	8.0	--	99.50	no	280	--	60
2,4,6-trichloroaniline	10000.0	2.5	--	99.97	no	320	--	120
2,4,6-trichlorophenol	--	--	--	99.90	no	285	--	120
2,4,6-trinitrophenol	--	--	--	99.90	no	285	--	120
2,4-dichloroaniline	259.0	0.5	--	99.80	no	275	--	60
2,4-dichlorophenol	500.0	2.0	--	99.60	no	320	--	60
2,4-dichlorophenol	23.5	0.4	--	98.40	no	280	--	60
2,4-dimethylphenol	8220.0	0.1	--	99.99	no	275	--	60
2,4-dinitrotoluene	10000.0	26.0	--	99.74	no	275	--	60
2,4-dinitrotoluene	10000.0	12.0	--	99.88	no	320	--	60

(continued)

Table 4.2 (continued)

COMPONENT	INFLUENT CONC. (MG/L)	EFFLUENT CONC. (MG/L)	FLOW RATE (GPM)	PERCENT REMOVAL	CATALYST REQUIRED	TEMP (C°)	PRESSURE (PSIG)	TIME (MIN)
2,6-dichlorobenzonitrile	498.0	39.0	--	92.10	no	280	--	60
2-(naphthoxy)-n,n-diethyl propionamide	362.0	0.9	--	99.80	no	280	--	60
2-chlorophenol	12410.0	15.0	--	99.88	yes	275	--	60
2-ethoxy ethanol	20000.0	200.0	--	99.00	yes	320	--	--
2-propanol	2230.0	40.0	--	98.20	no	280	--	60
2-propanol	2230.0	30.0	--	98.70	no	320	--	60
2-propanol	9900.0	170.0	--	98.30	no	280	--	60
3-nitrophenol	--	--	--	99.90	no	285	--	120
4-nitrophenol	10000.0	40.0	--	99.60	no	275	--	60
4-nitrophenol	10000.0	4.0	--	99.96	no	320	--	60
5,5-dimethylhydantoin	2254.0	16.0	--	99.30	no	280	--	60
acenaphthene	7000.0	0.5	--	99.99	no	275	--	60
acenaphthene	7000.0	2.8	--	99.96	no	320	--	60
acetaldehyde	20000.0	20.0	--	99.00	yes	320	--	--
acetic acid	20000.0	4000.0	--	80.00	yes	320	--	--
acetone	1680.0	10.0	--	99.40	no	280	--	60
acetone	1680.0	10.0	--	99.40	no	320	--	60
acetonitrile	1040.0	17.0	--	98.40	yes	275	--	60
acrolein	8410.0	3.0	--	99.96	no	275	--	60
acrolein	8400.0	3.4	--	99.96	no	320	--	60
acrolein	20000.0	200.0	--	99.00	yes	320	--	--
acrylonitrile	8060.0	80.0	--	99.00	no	275	--	60
acrylonitrile	8060.0	40.0	--	99.50	yes	275	--	60
acrylonitrile	8060.0	7.3	--	99.91	no	320	--	60
acrylonitrile	20000.0	200.0	--	99.00	yes	320	--	--

(continued)

Table 4.2 (continued)

COMPONENT	INFLUENT CONC. (MG/L)	EFFLUENT CONC. (MG/L)	FLOW RATE (GPM)	PERCENT REMOVAL	CATALYST REQUIRED	TEMP (C°)	PRESSURE (PSIG)	TIME (MIN)
ammonia	20000.0	1000.0	--	95.00	yes	320	--	--
aniline	20000.0	200.0	--	99.00	yes	320	--	--
aroclor 1254	20000.0	7400.0	--	63.00	no	320	--	120
benzene	1780.0	17.8	--	99.00	yes	320	--	--
benzidine	400.0	4.0	--	99.00	yes	320	--	--
benzyl chloride	6.0	0.3	--	95.00	yes	320	--	--
carbaryl	30.0	0.6	--	98.00	no	281	--	182
carbon disulfide	2300.0	23.0	--	99.00	no	275	--	--
carbon tetrachloride	4330.0	12.0	--	99.70	no	275	--	60
carbon tetrachloride	1160.0	11.6	--	99.00	yes	320	--	--
chlorobenzene	5535.0	1550.0	--	72.00	yes	275	--	60
chlorobenzene	500.0	25.0	--	95.00	yes	320	--	--
chloroform	4450.0	3.0	--	99.90	no	275	--	60
chloroform	8000.0	80.0	--	99.00	yes	320	--	--
chloroprene	960.0	96.0	--	99.00	yes	320	--	--
cresol	188.0	0.2	6.30	99.90	yes	279	1558	69
cyanide	309.0	6.0	6.30	99.00	yes	279	1558	69
cyanide	900.0	0.1	145.00	99.90	no	250	1000	90
dibutylphthalate	5230.0	26.0	--	99.50	no	275	--	60
dichloromethane	20000.0	200.0	--	99.00	yes	320	--	--
dichloromethane	252.0	1.0	--	99.00	no	280	--	60
dimethylaniline	1300.0	1.6	--	99.90	no	280	--	60
dinitrobenzenamine	--	--	--	90.60	no	285	--	120
dinitrophenol	--	--	--	99.90	no	285	--	120
dinoseb	37.1	0.2	--	99.50	no	281	--	182

(continued)

Table 4.2 (continued)

COMPONENT	INFLUENT CONC. (MG/L)	EFFLUENT CONC. (MG/L)	FLOW RATE (GPM)	PERCENT REMOVAL	CATALYST REQUIRED	TEMP (C°)	PRESSURE (PSIG)	TIME (MIN)
dipropylformamide	219.0	1.0	--	99.50	no	250	--	60
dipropylformamide	484.0	5.0	0.07	99.00	no	243	1582	130
dipropylformamide	484.0	20.0	0.07	95.90	no	224	1580	131
epichlorohydrin	20000.0	200.0	--	99.00	yes	320	--	--
ethanol	1530.0	60.0	--	96.10	no	280	--	60
ethanol	1530.0	10.0	--	99.30	no	320	--	60
ethanol	2800.0	1200.0	--	57.20	no	280	--	60
ethyl acetate	20000.0	200.0	--	99.00	yes	320	--	--
ethylene diamine	20000.0	200.0	--	99.00	yes	320	--	--
ethylene dibromide	4300.0	43.0	--	99.00	yes	320	--	--
ethylene dichloride	8000.0	80.0	--	99.00	yes	320	--	--
ethylene oxide	1000.0	10.0	--	99.00	yes	320	--	--
ethylenimine	20000.0	200.0	--	99.00	yes	320	--	--
ethylmercaptan	15000.0	1500.0	--	99.00	yes	320	--	--
formaldehyde	20000.0	200.0	--	99.00	yes	320	--	--
formic acid	25000.0	410.0	--	98.30	no	300	--	30
freon tf	3000.0	2.0	0.04	99.90	no	314	1943	128
hexachlorocyclopentadiene	100.0	1.0	--	99.00	yes	320	--	--
hexachlorocylcopentadiene	10000.0	15.0	--	99.90	no	300	--	60
hexane	13.0	0.1	--	99.00	yes	320	--	--
hexanone	17000.0	170.0	--	99.00	yes	320	--	--
hydrazine	1000.0	10.0	--	99.00	yes	320	--	--
hydrazine	12700.0	0.1	--	99.99	no	271	--	117
hydrogen cyanide	20000.0	200.0	--	99.00	yes	320	--	--
hydrogen sulfide	3000.0	30.0	--	99.00	yes	320	--	--

(continued

Table 4.2 (continued)

COMPONENT	INFLUENT CONC. (MG/L)	EFFLUENT CONC. (MG/L)	FLOW RATE (GPM)	PERCENT REMOVAL	CATALYST REQUIRED	TEMP (C°)	PRESSURE (PSIG)	TIME (MIN)
isophorone	4650.0	29.0	--	99.40	no	275	--	60
isopropyl alcohol	1700.0	400.0	0.04	76.50	no	314	1943	128
kepone	1000.0	690.0	--	31.00	no	280	--	60
malathion	11800.0	18.0	--	99.85	no	250	--	60
malathion	93.1	0.1	--	99.90	no	281	--	182
maleic anhydride	20000.0	200.0	--	99.00	no	270	--	--
mek	6000.0	1.0	0.04	99.90	no	314	1943	128
mek	6000.0	60.0	--	99.00	yes	320	--	--
mek	276.0	1.0	--	99.60	no	280	--	60
mek	276.0	3.0	--	98.90	no	320	--	60
mek	8200.0	1.0	--	99.98	no	280	--	60
mek	6000.0	1.0	--	99.90	no	314	1943	128
mercaptans	238.0	0.2	--	99.90	no	280	--	60
mercaptans	151.0	1.0	--	99.30	no	225	--	60
methanol	3230.0	1380.0	--	57.30	no	280	--	60
methanol	3230.0	290.0	--	91.00	no	320	--	60
methoxychlor	8.8	0.0	--	99.80	no	281	--	182
methyl mercaptan	20000.0	200.0	--	99.00	yes	320	--	--
methylene chloride	60.0	0.6	0.04	99.90	no	314	1943	128
n-nitrosodimethylamine	5030.0	22.0	--	99.60	no	275	--	60
n-nitrosodimethylamine	510.0	1.0	--	99.80	no	320	--	60
n-nitrosodimethylamine	20000.0	200.0	--	99.00	yes	320	--	--
naphthalene	30.0	0.3	--	99.00	yes	320	--	--
nitrobenzene	5125.0	255.0	--	95.00	yes	320	--	120
nitrobenzene	1900.0	95.0	--	95.00	yes	320	--	--

(continued)

Table 4.2 (continued)

COMPONENT	INFLUENT CONC. (MG/L)	EFFLUENT CONC. (MG/L)	FLOW RATE (GPM)	PERCENT REMOVAL	CATALYST REQUIRED	TEMP (C°)	PRESSURE (PSIG)	TIME (MIN)
organophosphorus	8320.0	490.0	--	94.00	no	300	--	60
p-dichlorobenzene	49.0	2.5	--	95.00	yes	320	--	--
pentachlorophenol	5000.0	135.0	--	97.30	yes	275	--	60
pentachlorophenol	5000.0	6.0	--	99.88	no	320	--	60
pentachlorophenol	14.0	0.7	--	95.00	yes	320	--	--
perchloroethylene	4000.0	0.9	0.04	99.90	no	314	1943	128
phenol	10000.0	20.0	--	99.80	no	275	--	60
phenol	10000.0	3.0	--	99.97	no	320	--	60
phenol	20000.0	200.0	--	99.00	yes	320	--	--
phthalimide	13000.0	132.0	--	99.00	no	225	--	60
propionitrile	391.0	7.0	--	98.20	yes	275	--	60
pyrene	500.0	0.3	--	99.95	no	275	--	60
pyridine	3910.0	570.0	--	85.40	yes	320	--	120
quinoline	38.0	13.0	0.10	65.80	yes	280	1558	69
styrene	300.0	3.0	--	99.00	yes	320	--	--
sulfide	6000.0	0.8	22.00	99.90	no	200	500	60
sulfide	9000.0	0.1	18.00	99.90	no	320	1900	50
tetrachloroethane	2900.0	29.0	--	99.00	yes	320	--	--
tetrachloroethylene	150.0	1.5	--	99.00	yes	320	--	--
tetrahydrofuran	20000.0	200.0	--	99.00	yes	320	--	--
toluene	4330.0	12.0	--	99.70	no	275	--	60
toluene	500.0	5.0	--	99.00	yes	320	--	60
toluene	80.0	1.0	--	98.80	no	280	--	60
toluene	80.0	1.0	--	98.80	no	320	--	--
toluene	30.0	0.5	0.04	98.30	no	314	1943	128

(continued)

Table 4.2 (continued)

COMPONENT	INFLUENT CONC. (MG/L)	EFFLUENT CONC. (MG/L)	FLOW RATE (GPM)	PERCENT REMOVAL	CATALYST REQUIRED	TEMP (C°)	PRESSURE (PSIG)	TIME (MIN)
toluene diisocyanate	10.0	0.1	--	99.00	yes	320	--	--
trichloroethylene	500.0	1.7	--	99.70	no	320	--	60
trichloroethylene	1000.0	10.0	--	99.00	yes	320	--	--
trichloroethylene	300.0	2.0	--	99.30	no	280	--	60
xylene	8385.0	20.0	0.04	99.80	no	314	1943	128
xylene	170.0	1.7	--	99.00	yes	320	--	--

Table 4.4. Waste Streams Appropriate for the WAO Process

Type of Waste
Municipal sludge
Night soil
Carbon regeneration
Acrylonitrile
Metallurgical coking
Petrochemical
Paper filler
Industrial activated sludge
Pulping liquor
Hazardous waste
Paper mill sludge
Explosives, 3,4,5-T, malathion
Monosodium glutamate
Polysulfide rubber
Textile sludge
Chrome tannery waste
Petroleum refining
Miscellaneous industrial sludges

Source: [1]

efficiency) ; 1-3% (grade=B, medium efficiency) ; 3-5% (grade=C, low efficiency). While such precise dividing points between these categories are not warranted, the data suggest that conversion efficiencies tend to decrease for more concentrated wastes and WAO is seldom applied to wastes containing more than 5% oxidizable compounds.

The types of waste streams which have been treated by the WAO process are given in Table 4.3. Both Table 4.1 and Table 4.3 reveal the wide range of applications for which the WAO process is suitable. Tables 4.4 and 4.5 show the results of full-scale WAO treatments of spent caustic and cyanide wastewaters, respectively, at the Casmalia Resources facility [1].

The efficiency of removal is affected by operating conditions. The most important WAO parameters are temperature, pressure, residence time, influent concentration of contaminant, and

Table 4.4. Wet Air Oxidation of Gulf Oil Spent Caustic Wastewater at Casmalia Resources

Operating conditions: Oxidation temperature, 515 °F (268 °C); Nominal residence time, 113 min; Waste flow rate, 5.3 GPM; Compressed air flow rate, 190 SCFM; Reactor pressure, 1610 psig; Residual oxygen concentration, 3.7%

Component Description	Influent Conc.	Effluent Conc.	% Reduction
COD, g/l	108.1	11.6	89.3
Total phenols, mg/l	15510	36	99.8
Total sulfur, mg/l	3580	3090	13.7
Sulfate sulfur, mg/l	570	2910	-
Organic sulfur [a], mg/l	3010	180	94.0
Sulfide sulfur, mg/l	<1.0	<1.0	-
pH	13.0	8.3	-
Total solids, g/l	88.6	59.7	32.6
Total ash, g/l	57.1	50.2	12.1
Volatile solids, g/l	31.5	9.5	69.8
Dissolved oxy. conc., mg/l	-	3680	-
Soluble chloride, mg/l	1510	550	63.6
Soluble fluoride, mg/l	4.4	1.3	70.5

[a] Organic sulfur = Total sulfur minus sulfate sulfur
Source: [1]

Table 4.5. Wet Air Oxidation of Cyanide Wastewater at Casmalia Resources

Operating conditions: Oxidation temperature, 495 °F (257 °C); Nominal residence time, 80 min; Waste flow rate, 7.5 GPM; Compressed air flow rate, 190 SCFM; Reactor pressure, 1220 psig; Residual oxygen concentration, 7.1%

Component Description	Influent Conc.	Effluent Conc.	% Reduction
COD, g/l	37.4	4.2	88.8
Cyanide, mg/l	25390	82	99.7
pH	12.6	9.0	-
Total solids, g/l	135.3	91.2	32.6
Total ash, g/l	112.9	77.4	31.4
Volatile solids, g/l	22.4	13.8	38.4
BOD_5, mg/l	-	603	-
Dissolved oxy. conc., mg/l	14710	1710	88.4
Soluble chloride, mg/l	-	773	-
Soluble fluoride, mg/l	30	29	3.3

[a] Organic sulfur = Total sulfur minus sulfate sulfur
Source: [1]

The high pressures are to force oxygen into solution,

Table 4.6. Toxic Contaminants Relatively Resistant to Wet Air Oxidation*

1. Halogenated condensed ring compounds such as pesticides
2. Halogenated aromatic compounds without other non-halogen functional groups e.g., PCB and chlorobenzene
3. Low molecular weight compounds

* without catalyst

Source: [13]

offgas oxygen concentration. If oxygen in the offgas is depleted, the reactor will lack oxygen, resulting in lower removal efficiencies. In general, removal efficiency increases with increasing temperature, pressure, residence time and influent oxygen concentration, and decreases with increasing influent waste concentration. Table 4.2 shows the results of several different tests on the same compound under different operating conditions. While the effects of these changes may seem small, these changes can radically affect the amount of unconverted contaminant in the process effluent. For example, decreasing removal efficiency from 99% to 98% doubles the effluent concentration of the contaminant. There may also be more subtle effects on the biodegradability of the effluent which are related to the degree to which the contaminant is oxidized in those cases where complete conversion to carbon dioxide and water, etc. does not occur. The effects of operating parameters are described in more detail below.

Certain types of organic compounds are not readily degraded under the conditions usually encountered in wet air oxidation. Table 4.6 lists the types of toxic contaminants which have been found to be resistant to WAO without the addition of a catalyst. The reason for the great resistance of the halogenated aromatics to the conventional WAO process may be the great strength of the halogen/ring bond in these compounds. The use of catalyst ions has been studied and the result shows increased oxidation rate for halogenated aromatics.

4.2.2. Catalytic Wet Air Oxidation

The use of a catalyst in the WAO process is essential in order to oxidize some compounds. When temperatures above 320°C are required to achieve the necessary oxidation or when an increase in temperature can not increase the percent of oxidation, use of a catalyst may be appropriate. Adding a catalyst will lower the activation energy, thus increasing the oxidation rate. Table 4.7 lists some compounds for which a catalyst can significantly improve the reaction rate. Although the operating temperatures in Table 4.7 for oxidation with and without catalyst were not the same, the last two entries (Aroclor 1254 and Pentachlorophenol) in Table 4.7 indicate that even though the temperature of the catalytic oxidation is lower than that of the non-catalytic case, the reduction efficiency for the catalytic case may be higher.

The copper ion is the most common catalyst, although bromide and nitrate catalysts have been employed. Also mentioned in one report were two proprietary catalysts used to increase oxidation of pesticides [4]. While addition of a copper catalyst may only increase the percent removal slightly, it may significantly reduce effluent toxicity as noted above. Depending on how hazardous or toxic the waste stream is, this reduction may be extremely important.

Because the catalyst is in the liquid phase, the regeneration and removal of the catalyst ions must occur downstream from the reactor. Due to pollution problems resulting from the use of a catalyst, the ions must be non-toxic or 100% recoverable from the liquid stream in order for their addition to be acceptable. Note also that if a catalyst is employed, the reactor, heat exchanger, and piping system must be resistant to corrosion caused by the catalyst as well as by the particular waste. This could increase the cost of the process.

Zimpro Inc. has only employed a catalyst in laboratory studies, and has yet to include it in a full-scale process. The companies operating wet air oxidation units may have employed a catalyst but no such reports have been found in the literature.

Table 4.7. Improved Compound Reductions When Using Catalyst

Compound	% reduction w/o catalyst T=275°C	% reduction w catalyst T=320°C
Benzylchloride	50	95
Chlorobenzene	50	95
p-Dichlorobenzene	20	95
Nitrobenzene	50	95
Acetic Acid	20	80
Ammonia	0	95
	T=320°C	T=275°C
Aroclor 1254	63	99.62
Pentachlorophenol	81.96	99.997

Source: [4]

4.2.3. Important Operating Parameters

The efficiency of the WAO process is controlled by several parameters which influence the rate of oxidation and the time available for reaction. These operating parameters are the temperature and pressure in the reactor, the residence time of the waste in the reactor, and the oxygen concentration exiting the reactor [3]. The influent concentration of contaminant, while generally not adjustable by the operator, has a significant impact on effluent quality.

Elevated temperature is essential for the oxidation of waste. Oxidation rates generally increase exponentially with temperature. Depending on the compounds to be oxidized, the operating temperature may range from 175 to 325 °C. To obtain a specified removal, the temperature chosen will also depend on the chemical oxygen demand entering the reactor, the pressure, the residence time and the unreacted oxygen in the effluent. If the appropriate temperature is not obtained, oxidation rates will be unacceptably low.

If a catalyst is added, the activation energy will be lowered, and hence, the necessary reaction rate can be achieved at a lower temperature. Several tests were made on an acrylonitrile waste in order to find the relationship between temperature and conversion [5]. At 275°C, 99% conversion was obtained without catalyst. At 275°C with the use of a catalyst, the conversion was 99.5%. However, at 329°C, without the use of a catalyst, 99.91% conversion was obtained. If a catalyst were applied to the last run, the percentage would presumably have increased. However, an economic decision must be made concerning temperature and use of catalyst.

The time spent in the reactor affects the amount of oxidation occuring in the waste. The residence time is dependent upon the flow rate of waste entering the system as well as the length and diameter of the tubular flow reactor. Typical residence times are 60 to 120 min. The oxidation of a particular packet of waste is the greatest in the first 30 min of travel through the column, and then it slowly decreases as the packet travels toward the exit of the column. Residence time will decrease if flow rate increases. Therefore, if flow rates increase, conversion will decrease. In designing a process to achieve a specified conversion, it is usually more economical to increase temperature than to increase the residence time by building a larger reactor.

The WAO process is a liquid phase process at normal operating temperatures and pressures. At such temperatures and pressures (which are near the critical condition for water), water in the liquid phase allows for greater solubility of oxygen and also provides an excellent heat capacity to buffer temperature changes. The process pressure is maintained between 300-3000 psig to prevent excessive evaporation of the liquid. The pressure of the exit stream is released after leaving the cooler [4].

The amount of oxygen supplied also contributes to the amount of oxidation which will occur. The greater the amount of O_2, the greater the oxidation rate. Normally 15,000 mg/L of O_2 is sufficient for the WAO process, but as much as 70,000 mg/L [5] has been used to ensure plentiful supply of oxygen for oxidation. If the COD of the waste-stream is greater than the amount of oxygen supplied, the reaction is starved for oxygen, and the rate decreases as well as the percentage of waste oxidized. Samples of reactor effluent are usually taken to determine how much unreacted dissolved oxygen remains.

4.2.4. Construction Materials of WAO Process

Materials for the construction of the WAO process are dependent upon the following factors: the type of waste entering the system, whether a catalyst is present, the amount being processed, and the design life of the process. The materials established as appropriate for WAO are stainless steel, nickel, and titanium alloys; the latter being used for the extremely corrosive wastes containing heavy metals. The parts of the structure which need protection are the heat exchanger, reactor, pipes, and cooling system.

Grade 12 titanium is a type of titanium alloy widely used due to its corrosion resistance and lower price compared to other types of titanium alloys [6]. It is appropriate for wastes with a pH ranging from 1-12. Stainless steels are used for wastes containing sulfates. Other alloys and materials are listed in Table 4.8.

Table 4.8. Materials That May Be Applicable for WAO Construction *

Material	Elemental Characteristics	Characteristics
Hastelloy G	30%-50% Ni Cr, Mo, Cu	Good for acid solution
Super SS	Ni, Cr	
Hastelloy C	Ni, Cr, Mo	
Grade 7	Ti, Pd	
Glass-lined steel		Not resistant to HF. Inert to most substances
Tantulum	Ta	Not resistant to bromine in combination with primary alcohols. Not resistant to sulfur trioxide or HF.
Tantulum alloy	Ta, Cm	Alloy used for cost reduction purposes.

* All materials listed are resistant to chlorides when oxidized.

Source: [14]

4.3. COST OF THE PROCESS

The total cost of the WAO process includes various costs associated with the construction and operation of the plant. Figures 4.2 and 4.3 are graphs showing the capital, and operating and maintenance costs respectively as supplied by Zimpro [3]. The capital cost is dictated by the type of waste, the capacity of the system, the required oxygen demand reduction, the treatment objectives, and the materials for construction. Zimpro's capital cost estimates are based on the following operating parameters: temperature = 280° C, pressure = 2000 psig, and the COD reduction = 50,000 mg/L. The design residence time is not specified. The cost of building and foundations are not included.

The operating and maintenance costs include maintenance, labor, and utility costs. Figure 4.3 shows the operating and maintenance costs as well as the utility and labor costs as a function of flow rate. The cost of electricity and cooling water are assumed to be \$0.05/kwh and \$0.25/1000 gal respectively.

Neither the capital nor the operating costs include downstream treatment, e.g. biological treatment, of the liquid effluent which is frequently needed. However, an off-gas scrubber is included in the cost estimates from Zimpro. All cost estimates are based on costs for existing WAO process units which have throughput varying from 6 to 3,000 gpm. The nominal plant life is 20 years, with occasional repairs done on the equipment.

Figures 4.4 and 4.5 are the annual capital costs, amortized over 10 and 20 years, respectively, for flow rates between 6 and 70 gpm. Costs for any unit designed for a flow rate less than 6 gpm, are based on those for a 6 gpm installation (\$ 220,000 for 10 years and \$160,000 for 20 years). The amortized annual capital cost data in Figures 4.4 and 4.5 were calculated from the following capital recovery equation using the average capital costs (within the shaded band) at various flow rates in Figure 4.2 and assuming 10% annual interest rate

$$\text{Annual Capital Cost} = \text{Capital Cost} \times [i\,(i+1)^n / (i+1)^n - 1] \qquad (4.1)$$

where i=0.1 (10%), n= # of yrs. (10 and 20 yrs).

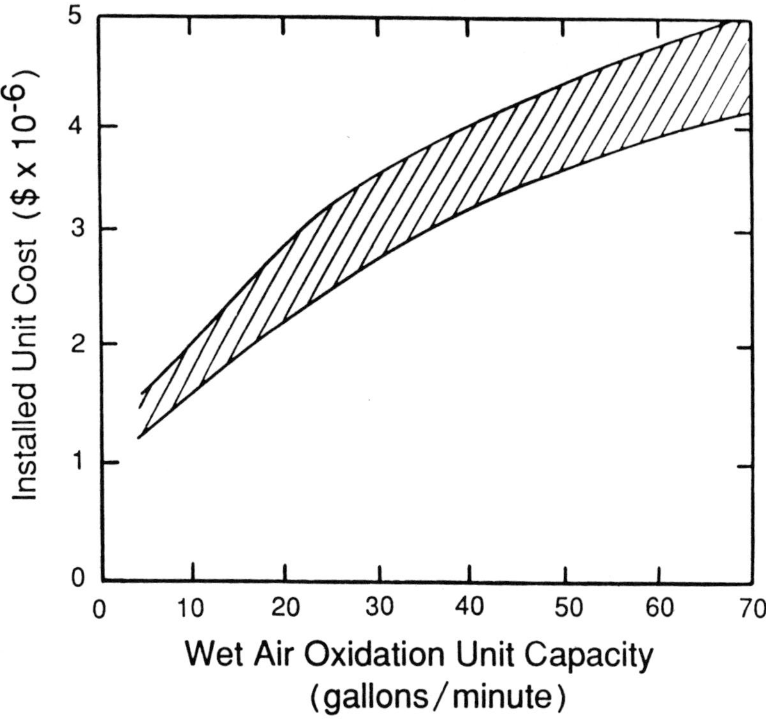

Figure 4.2 Capital, Operating and Maintenance Costs.

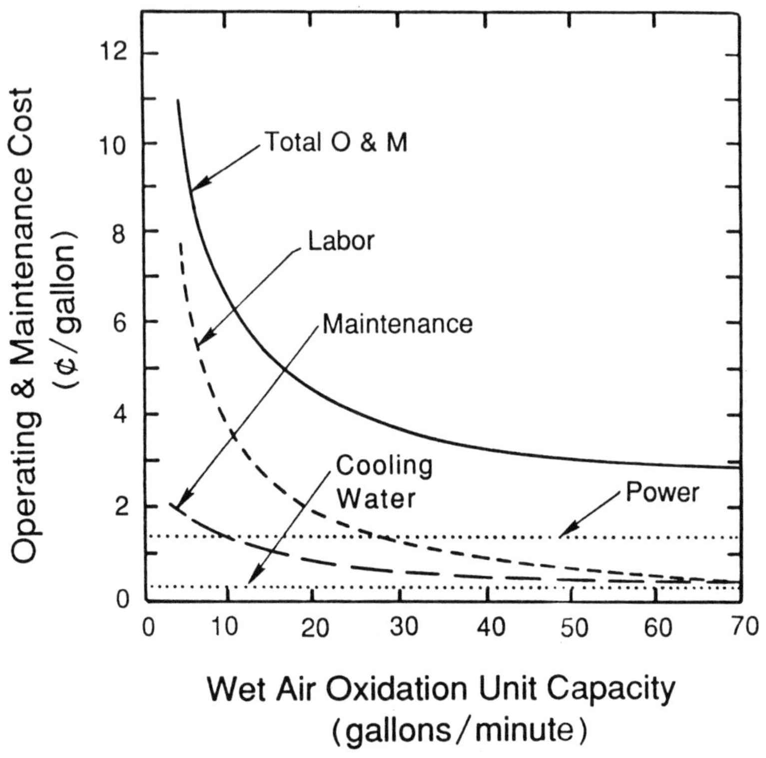

Figure 4.3 Operating, Maintenance, Utility and Labor Costs.

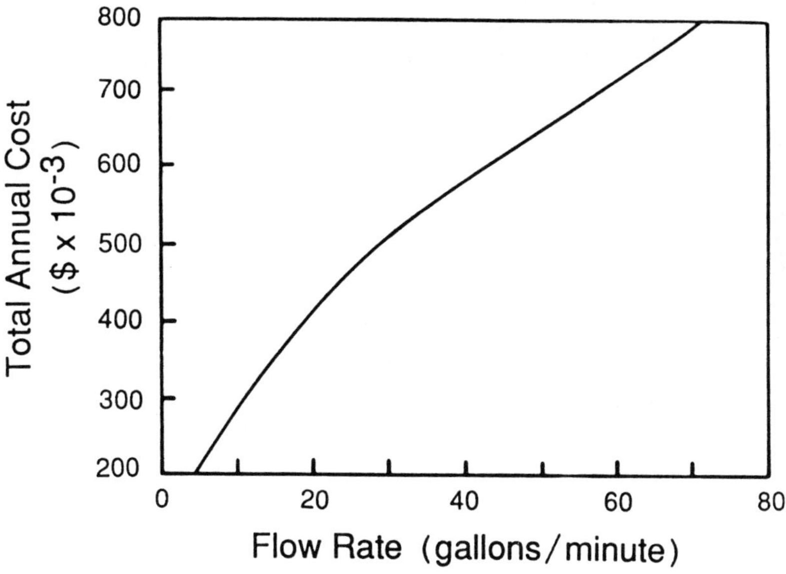

Figure 4.4 Annual Capital Cost for 10 Year Amortization.

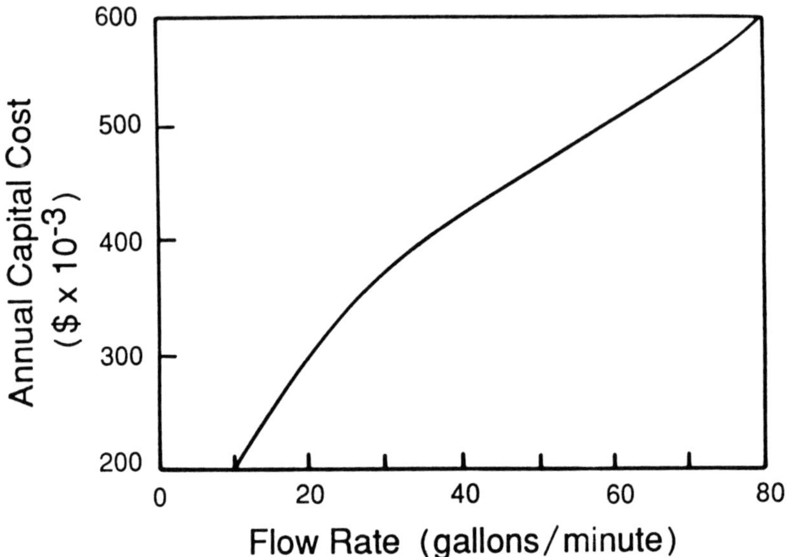

Figure 4.5 Annual Capital Cost for 20 Year Amortization.

The annual operating and maintenance costs are given in Figure 4.6. They were obtained by multiplying the figure in the O & M curve in Figure 4.3 with the annual volume (calculated from the flow rate using 8 working hours/day and 250 working days per year). The total annual cost of operating a WAO plant is the sum of the annual capital cost and annual operating and maintenance (O & M) cost. Figures 4.7 and 4.8 show the total annual costs of WAO facilities (for flow rates greater than 6 gpm) having 10 and 20 years of plant life, respectively.

In some cases the WAO process is used to create a product which is saleable on the market. An example of this is the steel industry, and their disposal of coke oven-gas [7]. Off-gas with a 90% sulfur content enters the WAO system and is converted in the presence of ammonium thiocyanate to ammonium sulfate and sufuric acid. The ammonium sulfate is evaporated to crystals and is sold as fertilizer. The sulfuric acid is used to reduce the acid consumption in the mill itself.

4.4 ASSESSING THE TREATABILITY OF CALIFORNIA WASTES BY THE WAO PROCESS

A database manipulation program was written to determine which of the wastestreams in the 1985 California BGR database [8] are suitable for treatment using wet air oxidation and to estimate the cost of such a treatment. The flow chart for the program to assess treatability and estimate cost is given in Figure 4.9 and the results are shown in Table 4.9. A description of the program and the information generated from it is discussed below.

4.4.1. Results on Treatability Using WAO

Each waste stream in the BGR database was studied to assess the suitability of WAO for treating the stream. Any stream described as a solid was rejected as unsuitable, as were streams containing less than 85% water. For liquid streams with 85% water or more, each component was compared with the list of compounds treatable by WAO as given in Table 4.2. If one or more components in the wastestream was on the list of treatable compounds in Table 4.2, and the sum of those concentrations fell between the three previously mentioned concentration limits (0-1%, 1-3%, 3-5%), the wet air oxidation process was deemed appropriate with corresponding grades

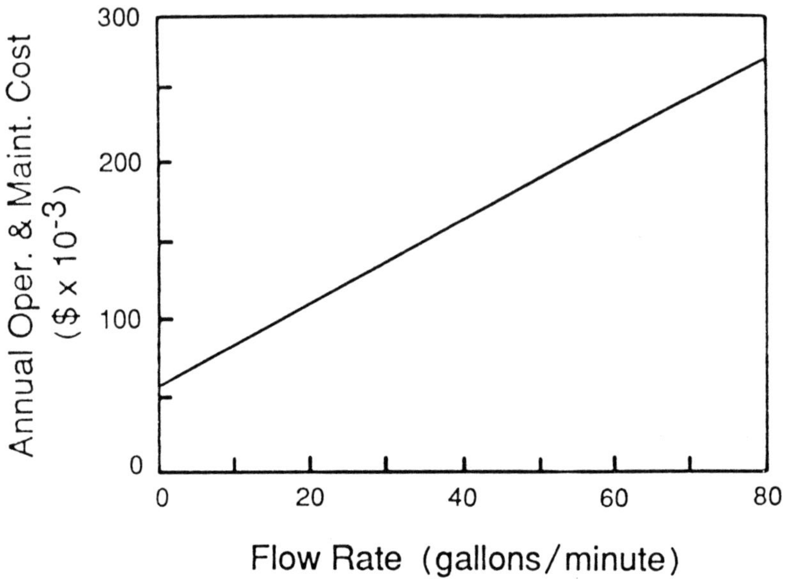

Figure 4.6 Annual Operating and Maintenance Costs.

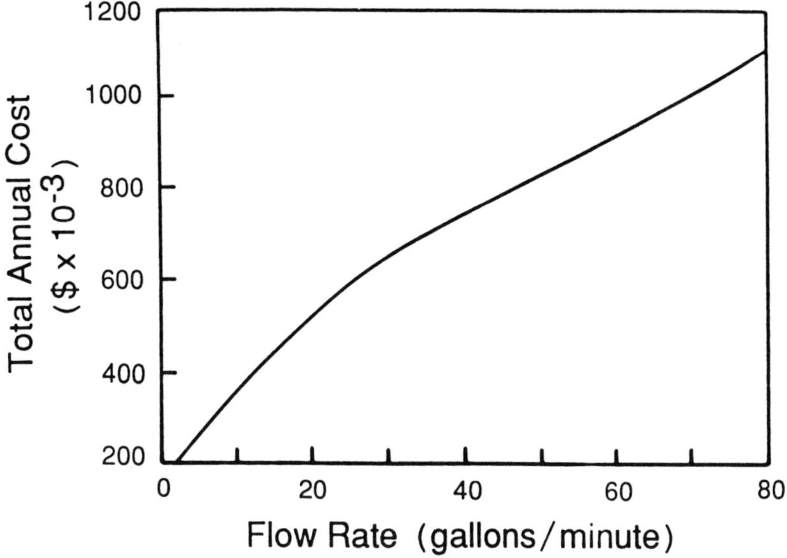

Figure 4.7 Total Annual Cost for 10 Year Amortization.

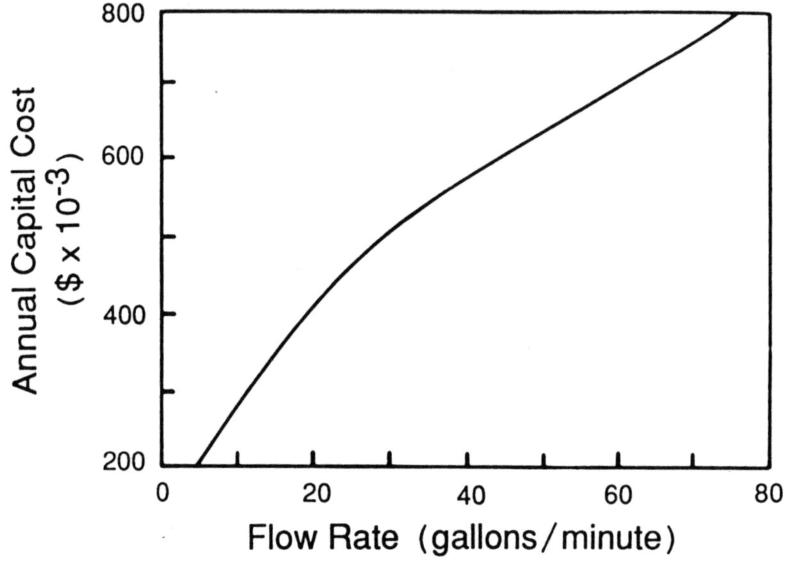

Figure 4.8 Total Annual Cost for 20 Year Amortization.

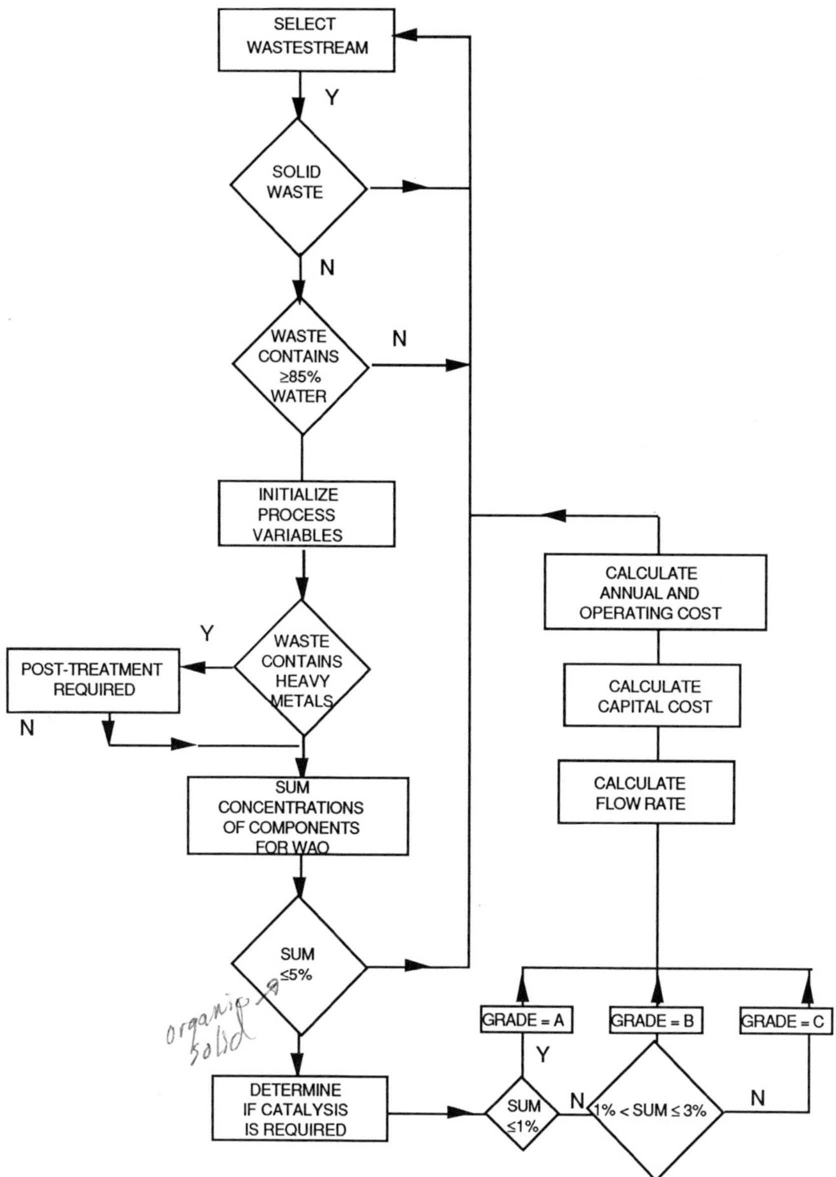

Figure 4.9 Flow Diagram for Selection and Cost Estimation of Wet-Air Oxidation Processes.

Table 4.9 Typical Wastestreams Treatable by WAO

SIC	VOLUME (TONS)	CAPITAL COST($)	ANNUAL OPERATING COST (10 YR)	COST (20 YR)	CATALYST REQUIRED	% WATER	GRADE	DESCRIPTION OF GENERATING PROCESS
37	148.49	1,400,000	276,853	216,853	yes	100.00	A	machining of rocket engine parts
28	23.00	1,400,000	276,217	216,217		99.50	A	from manufacturing activity of organic pesticide chemicals
34	12.50	1,400,000	276,163	216,163	yes	100.00	A	process sewer line
34	33.33	1,400,000	276,269	216,269	yes	99.70	A	precious metal finishing process containing cyanides
35	54.38	1,400,000	276,376	216,376		100.00	A	34,35 monitoring well sampling and testing
35	60.42	1,400,000	276,406	216,406		100.00	A	34,35 monitoring wells pumping, testing
36	0.01	1,400,000	276,100	216,100		99.50	A	electronic components
99	0.46	1,400,000	276,102	216,102		100.00	A	mixed solvent/cleanup material
51	583.33	1,400,000	279061	219,061	yes	99.90	A	contaminated rainwater
36	0.06	1,400,000	276,100	216,100	yes	99.90	A	circuit board manufacturing
34	14.20	1,400,000	276,172	216,172		99.90	A	metal stripping
37	1761.00	1,400,000	285056	225,056		99.90	A	waste water treatment facility for coating operations
34	2.08	1,400,000	276,111	216,111	yes	99.80	A	precious metals finishing process containing cyanides
28	162.20	1,400,000	276,923	216,923		99.80	A	under water clean-up
34	6.25	1,400,000	276,132	216,132	yes	99.90	A	precious metals finishing process containing cyanides
28	3237.50	1,406,249	301,192	237,499		99.70	A	equipment cleaning
37	1.25	1,400,000	276,106	216,106		100.00	A	clarifier sludge
37	1.12	1,400,000	276,106	216,106	yes	99.50	A	filter press cake from waste water pretreatment
97	2841.67	1,400,000	290,576	230,576	yes	99.90	A	chemical toilet cleanout
34	2.08	1,400,000	276,111	216,111	yes	99.80	A	precious metal finishing process
36	0.05	1,400,000	276,100	216,100	yes	99.70	A	electronics assembly

(continued)

Table 4.9 (continued)

SIC	VOLUME (TONS)	CAPITAL COST($)	ANNUAL OPERATING COST (10 YR)	COST (20 YR)	CATALYST REQUIRED	% WATER	GRADE	DESCRIPTION OF GENERATING PROCESS
36	11.10	1,400,000	276,156	216,156	yes	100.00	A	(cu) spill
48	5.00	1,400,000	276,125	216,125		99.90	A	pumpings from manholes
34	20.83	1,400,000	276,206	216,206	yes	99.80	A	precious metals finishing process containing cyanides.
37	250.00	1,400,000	277,368	217,368	yes	99.90	A	clean-up clarifiers serving water wash paint booths,steam cleaner
37	48.80	1,400,000	276,348	216,348		99.80	A	cleaning of waste tank
29	277.00	1,400,000	277,505	217505	yes	99.90	A	contaminated water cleanup
36	1.15	1,400,000	276,106	216,106	yes	99.90	A	semiconductor mfg.
28	12.50	1,400,000	276,163	216,163	yes	100.00	A	adhesive mixer wash
07	104.17	1,400,000	276,628	216,628		99.50	A	pesticide application,transportation
99	4.17	1,400,000	276,121	216,121	yes	99.90	A	
29	39.30	1,400,000	276,299	216,299		85.30	A	cleaning sludge from cooling tower
29	20.00	1,400,000	276,201	216,201	yes	87.30	A	clean up solution from ammonia tank
37	16.67	1,400,000	276,185	216,185		88.00	A	mfg. process
39	5.00	1,400,000	276,125	216,125	yes	88.00	A	electroless copper plating of printed circuit boards
36	3.00	1,400,000	276,115	216,115	yes	89.20	A	waste treatment plant sludge from plating operations
39	71.88	1,400,000	276,465	216,465	yes	89.20	A	rinsewater and plating solutions
29	200.00	1,400,000	277,115	217,115	yes	90.40	A	tank cleaning
36	0.70	1,400,000	276,104	216,104	yes	90.00	A	sludge from plating solution
38	143.60	1,400,000	276,828	216,828		89.90	A	waste oil from machines
34	83.33	1,400,000	276,523	216,523		90.00	A	degreaser sludges
36	2.00	1,400,000	276,110	216,110	yes	90.40	A	thin film head production
34	60.00	1,400,000	276,404	216,404	yes	89.80	A	pretreatment before sewering of metal finishing effluent
29	91.00	1,400,000	276,562	216,562		90.00	A	mfg of asphalt emulsions product samples,packing leaks
37	412.00	1,400,000	278,191	218,191		90.00	A	aircraft painting

(continued)

Table 4.9 (continued)

SIC	VOLUME (TONS)	CAPITAL COST($)	ANNUAL OPERATING COST (10 YR)	COST (20 YR)	CATALYST REQUIRED	% WATER	GRADE	DESCRIPTION OF GENERATING PROCESS
37	5.00	1,400,000	276,125	216,125		89.80	A	spent color chem film for aluminum mixed w/spent hcl
39	26.00	1,400,000	276,232	216,232	yes	90.00	A	
36	16.00	1,400,000	276,181	216,181	yes	91.10	A	printed circuit board manufacturing
34	5.00	1,400,000	276,125	216,125		91.50	A	cleaning of metal after forming
36	1.83	1,400,000	276,109	216,109	yes	93.00	A	printed circuit manufacturing process
33	313.00	1,400,000	277,688	217,688	yes	92.70	A	precious metal reclaim
36	32.00	11,400,000	276,262	216,262		93.20	A	machining
99	89.00	1,400,000	276,551	216,551		93.40	A	substance used for photographic processing
36	0.48	1,400,000	276,102	216,102	yes	93.90	A	electrical and electronic machinery
36	12.00	1400000	276,161	216,161	yes	94.00	A	clean-up of clarifier
36	899.17	1,400,000	280667	220667	yes	93.70	A	spent hydrofluoric acid solution used in mfg of semiconductor
36	0.50	1,400,000	276,103	216,103		94.30	A	photographic developing process
28	12.50	1,400,000	276,163	216,163	yes	94.00	A	washwaters from subdivision/packaging line
73	260.00	1,400,000	277,419	217,419		93.90	A	subtance used for photographic processing
42	71.47	1,400,000	276,462	216,462		94.40	A	interior cleaning of bulk liquid highway transportation equip.
39	12.00	1,400,000	276,161	216,161	yes	95.20	A	deburring of aluminum castings
36	12.00	1,400,000	276,161	216,161	yes	94.80	A	clarifier sludge
37	17.50	1,400,000	276,189	216,189	yes	94.80	A	cleaning of waste tank
42	637.00	1,400,000	279,334	219,334		95.20	A	interior cleaning of bulk liquid highway transportation equip.
28	2252.00	1,400,000	287,561	227,561		96.20	A	manufacture of 10,10-oxybisphenoxarsine (obpa)
28	15145.00	3,145,739	644,948	503,058	yes	95.50	A	manufacture of nitroplasticizers
36	19.71	1,400,000	276,200	216,200	yes	96.00	A	copper etching solution
36	2.40	1,400,000	276,112	216,112	yes	95.90	A	acid cleaning of silicon wafers/spill kits where used

(continued)

Table 4.9 (continued)

SIC	VOLUME (TONS)	CAPITAL COST($)	ANNUAL OPERATING COST (10 YR)	COST (20 YR)	CATALYST REQUIRED	% WATER	GRADE	DESCRIPTION OF GENERATING PROCESS
29	341.00	1,400,000	277,830	217,830		96.00	A	chemical cleaning solution
34	0.25	1,400,000	276,101	216,101		97.00	A	metal plating
37	21.00	1,400,000	276,207	216,207	yes	97.00	A	manufacture of rocket chambers
99	5.00	1,400,000	276,125	216,125	yes	97.30	A	plating rinses/spent stripping solution
28	593.00	1,400,000	279,110	219,110	yes	96.50	A	oilfield chemicals (blending only)
38	79.00	1,400,000	276,501	216,501		97.10	A	3 water curtain spray paint booths
39	19.40	1,400,000	276,198	216,198	yes	97.30	A	clarifier
36	130.00	1,400,000	276759	216759	yes	96.50	A	etching silicon material with hydrofluoric acid
34	15.00	1,400,000	276,176	216,176	yes	97.10	A	electroplating
34	2.08	1,400,000	276,111	216,111		98.00	A	metal coating treatment
38	69.00	1,400,000	276,450	216,450	yes	98.00	A	p.c. fabrication, chromate anodize line
28	25.00	1,400,000	276227	216,227	yes	98.00	A	solution coating of synthetic substrates
28	150.00	1,400,000	276,861	216,861		98.00	A	equipment and floor cleaning wash water
99	0.62	1,400,000	276,103	216,103		98.20	A	metal cleaning/stripping
39	12.50	1,400,000	276163	216,163	yes	98.00	A	carbon adsorber condensate
36 discharge	41.67	1,400,000	276,311	216,311	yes	97.50	A	clarifier pump-out wastewater
36	1.25	1,400,000	276,106	216,106		98.20	A	generated plating bath solution from electroplating operations
36	1000.00	1,400,000	281,179	221,179	yes	98.00	A	spent liquid mixture from electroplating operations
99	1.04	1,400,000	276,105	216,105	yes	99.00	A	rainwater collected in drums previously containing trichloroeth.
34	0.04	1,400,000	276,100	216,100		99.00	A	metal plating
38	0.04	1,400,000	276,100	216,100	yes	99.00	A	chemical laboratory waste
25	21.00	1,400,000	276,207	216,207	yes	98.50	A	(cu) site cleanup material
07	260.87	1,400,000	277,424	217,424		99.20	A	plating shop operation
36	2.75	1,400,000	276,114	216,114		99.00	A	silver plating process for copper bus bar

(continued)

Table 4.9 (continued)

SIC	VOLUME (TONS)	CAPITAL COST($)	ANNUAL OPERATING COST (10 YR)	COST (20 YR)	CATALYST REQUIRED	% WATER	GRADE	DESCRIPTION OF GENERATING PROCESS
34	2.08	1,400,000	276,111	216,111	yes	98.50	A	spent silver plating solution
36	40.00	1,400,000	276,303	216,303		99.00	A	aluminum chemical conversion coating
37	100.83	1,400,000	276,611	216,611		99.00	A	fueling operation
49	92.40	1,400,000	276,569	216,569	yes	99.00	A	flushing contaminated tank
49	33.20	1,400,000	276,268	216,268	yes	98.50	A	tank flushing
49	32.70	1,400,000	276,266	216,266	yes	99.00	A	tank flushing
49	30.20	1,400,000	276,253	216,253	yes	99.00	A	tank flushing
49	10.40	1,400,000	276,153	216,153	yes	99.00	A	tank flushing
36	7.10	1,400,000	276136	216,136	yes	99.20	A	manufacture of printed circuit boards
36	0.51	1,400,000	276103	216,103		99.40	A	waste gold plating solution
36	2.57	1,400,000	276113	216,113	yes	98.60	A	spent plating bath solution
28	29.00	1,400,000	276247	216247		99.00	A	tank bottom sediment
28	2.00	1,400,000	276110	216,110	yes	99.00	A	tank bottom sediment
07	6.00	1,400,000	276130	216,130	yes	99.30	A	pesticide application
26	37.50	1,400,000	276290	216,290		85.00	B	chemical and pharmaceutical mfg.
34	37.50	1,400,000	276290	216,290	yes	86.10	B	spent cyanide - zinc plating solution
36	180.10	1,400,000	277,014	217,014	yes	85.90	B	semiconductor manufacturing
28	31.00	1,400,000	276257	216,257	yes	86.80	B	waste water from neutralization tank sludge
37	20.00	1,400,000	276,201	216,201		89.00	B	metal finishing activities
34	4.17	1,400,000	276,121	216,121		88.60	B	metal cleaning and plating
28	20.00	1,400,000	276,201	216,201	yes	90.00	B	mfg. of photographic and electronic chemicals
95	4.00	1,400,000	276,120	216,120	yes	90.00	B	solvent/water waste from office equipment repair shop
39	4.20	1,400,000	276,121	216,121		90.00	B	parts cleaning
28	395.83	1,400,000	278,109	218,109	yes	90.90	B	blending and packaging of consumer products for home use

(continued)

Table 4.9 (continued)

SIC	VOLUME (TONS)	CAPITAL COST($)	ANNUAL OPERATING COST (10 YR)	COST (20 YR)	CATALYST REQUIRED	% WATER	GRADE	DESCRIPTION OF GENERATING PROCESS
36	1.50	1,400,000	276,108	216,108		91.40	B	positive photoresist process
99	0.46	1,400,000	276,102	216,102		94.00	B	printed wiring board production
i36	1280.00	1,400,000	282,604	222,604	yes	94.60	B	rainwater collection-chemical vaults
37	52.00	1,400,000	276,364	216,364	yes	95.50	B	rinse water from resin impregnation of metal castings
36	2.10	1,400,000	276,111	216,111	yes	96.00	B	electronics assembly
28	2000.00	1,400,000	286,275	226,275	yes	96.50	B	washing of batching tanks, filling lines, and transfer lines
37	37.50	1,400,000	276,290	216,290	yes	96.70	B	plating
28	7066.67	2,100180	433,339	338,433		97.00	B	cleaning solvents
28	37.60	1,400,000	276,291	216,291		98.00	B	steam regeneration of spent activated carbon
36	45.83	1,400,000	276,332	216,332	yes	97.70	B	cleaning and stripping of parts
f37	6.00	1,400,000	276130	216,130		97.50	B	rinse water from organic paint stripper
97	1720.00	1,400,000	284,846	224,846	yes	98.30	B	oil/water separator bottom sludge from effluent to iwtp
38	0.04	1,400,000	276,100	216,100		97.50	B	gold stripping
99	13.33	1,400,000	276,168	216,168		97.80	B	silver recovery
39	0.04	1,400,000	276,100	216,100		98.00	B	disposal of obsolete stock
37	0.83	1,400,000	276,104	216,104	yes	98.20	B	cleaning solvent
36	10.00	1,400,000	276,151	216,151		97.50	B	electroplating of electronic parts
99	22.29	1,400,000	276,213	216,213	yes	98.70	B	chemical nuetralization and waste water
36	4.17	1,400,000	276,121	216,121		98.50	B	plating, absorbent pillows
33	16.67	1,400,000	276,185	216185	yes	98.80	B	process sludge
37	6.92	1,400,000	276135	216,135	yes	99.00	B	metal finishing operations
39	115.83	1,400,000	276,688	216,688	yes	98.50	B	holding tank for etch room wastes
20	20.00	1,400,000	276,201	216,201	yes	98.90	B	(cu) site clean-up
28	868.48	1,400,000	280,510	220,510		86.50	C	waste in manufacturing process

(continued)

Table 4.9 (continued)

SIC	VOLUME (TONS)	CAPITAL COST($)	ANNUAL OPERATING COST (10 YR)	COST (20 YR)	CATALYST REQUIRED	% WATER	GRADE	DESCRIPTION OF GENERATING PROCESS
34	6.25	1,400,000	276132	216,132		86.50	C	tin stripping
29	600.00	1,400,000	279,146	219,146		89.40	C	spent stretford solution
36	19.00	1,400,000	276,196	216,196	C	90.00	C	production of printed circuit boards
12	19.00	1,400,000	276,196	216,196	C	90.00	C	production of printed circuit boards
37	50.00	1,400,000	276,354	216,354	C	91.50	C	misc. alkaline waste from labs
37	14.20	1,400,000	276,172	216,172	YES	93.00	C	test operations
26	37.50	1,400,000	276,290	216,290		93.00	C	chemical and pharmaceutical manufacturing
34	0.56	1,400,000	276,103	216,103		95.00	C	gold strip
34	3.00	1,400,000	276115	216,115	yes	95.00	C	stripping gold solution
13	39.00	1,400,000	276,298	216298		95.00	C	crude oil production
99	2.08	1,400,000	276111	216,111	yes	95.00	C	printed wiring board production
37	10.00	1,400,000	276,151	216,151		95.20	C	paint stripping
28	14.00	1,400,000	276,171	216,171	yes	95.40	C	mfg. of photographic and electronic chemicals
97	1.50	1,400,000	276,108	216,108	yes	95.00	C	
36	10.00	1,400,000	276,151	216,151		95.00	C	metal finishing
36	2.50	1,400,000	276,113	216,113		96.00	C	plating

(grade A for high removal efficiency, grade B for medium removal efficiency and grade C for low removal efficiency). If a wastestream assessed as treatable by WAO contains component(s) that required catalyst for removal, the stream was listed as requiring catalyst addition. If the wastestream contained heavy metals, it was listed as requiring post-treatment.

The results of this treatability assessment must be interpreted carefully. The list of treatable components in Table 4.2 is certainly incomplete, representing only those substances for which reports were found indicating significant removal using WAO. There are surely many components not on this list which are suitable for treatment using WAO. Moreover, the computer algorithm neglected possible interference with WAO oxidation caused by presence of other components in the wastestream.

4.4.2. Cost Estimation for Wet Air Oxidation

After the waste streams suitable for WAO treatment were identified, the capital and annual operating costs were calculated from the literature values. Certain assumptions were made in order to calculate the capital and annual operating costs for a WAO unit for each wastestream. The assumptions that were used to compute the capital and annual operating costs are listed below.

1. Due to the fact that the reported cost data were a function of wastestream flow rate, the annual discharge volume as reported in BGR database had to be converted to the volumetric flow rate. We used 250 working days a year and 8 working hours a day as the basis for calculating the wastestream flow rate. Thus,

$$Q\text{ (gal/min)} = \frac{\text{annual vol.(tons/yr)x 240 (gal/ton)}}{\text{(250 day/yr)(8 hr/day)(60 min/hr)}}$$

$$= (\text{Vx}240)/(250\text{x}8\text{x}60) = \text{V}/250 \qquad (4.2)$$

where the density of water (240 gal/ton) is used as the conversion factor from tons to gallons.

2. If the calculated flow rate, Q, from equation (4.2) is less than or equal to the smallest flow rate (6 gpm in this case) for which cost data were found, the capital costs for the smallest flow rate was chosen.

3. The annual cost of WAO process is the sum of annual capital cost and operating and maintenance (O & M) costs. The annual capital costs were calculated using 10 and 20 years of life for the equipment and 10% annual interest rate . The formula for the annual capital cost is shown in Equation 4.1 in Section 4.3.

With the above assumptions, the capital and annual operating costs for wet air oxidation were computed using the flow sheet shown in Figure 4.9.

Capital Cost

The lowest flow rate shown in Figure 4.2 is 6 gal/min, for which the average capital cost (in the shaded region) is \$1.4 million. Therefore, according to assumption (2),

$$\text{Capital cost} = \$1.4\text{x}10^6 \quad \text{for } Q < 6 \text{ gal/min} \tag{4.3}$$

For Q > 6 gal/min, capital cost and flow rate were correlated by fitting the average data within the shaded area in Figure 4.2 to a third-degree polynomial. The correlation coefficient is 1.0, indicating that the fit is excellent. Thus, for Q > 6 gal/min,

$$\text{Capital cost} = 6.93\text{x}10^5 + 1.2\text{x}10^5\, Q - 1583.7\, Q^2 + 9.75\, Q^3 \tag{4.4}$$

Annual Operating Cost

The annual capital costs, amortized over 10 and 20 years, were calculated from Equation (4.1), using the appropriate capital cost as determined from Equation (4.3) or (4.4). For flow rates less than 6 gpm, the annual capital cost of a 6 gpm facility was used. Thus,

$$\text{Annual Capital Cost} = \$220{,}000 \quad \text{for } n=10 \text{ yrs, } Q < 6 \text{ gpm} \tag{4.5}$$

$$= \$160{,}000 \quad \text{for } n=20 \text{ yrs, } Q < 6 \text{ gpm} \tag{4.6}$$

For Q > 6 gpm, the correlated equations in Figures 4.4 and 4.5 were used

$$\text{Annual Capital Cost} = 1.13\text{x}10^5 + 1.945\text{x}10^4\,Q - 257.14\,Q^2 + 1.583\,Q^3 \quad \text{for } n=10,\ Q > 6 \text{ gpm} \tag{4.7}$$

$$= 8.14\text{x}10^4 + 1.41\text{x}10^4\,Q - 185.79\,Q^2 + 1.144\,Q^3 \quad \text{for } n=20,\ Q > 6 \text{ gpm} \tag{4.8}$$

The annual operating cost is the sum of annual capital cost and annual O & M cost. Using a quadratic fit to the annual O&M cost curve in Figure 4.7, we find

$$\text{Annual Op. Cost} = 2.761\text{x}10^5 + 2535.7\,Q + 2\,Q^2,\ n=10,\ Q < 6 \text{ gpm} \tag{4.9}$$

$$= 2.160\text{x}10^5 + 2535.7\,Q + 2\,Q^2,\ n=20,\ Q < 6 \text{ gpm} \tag{4.10}$$

$$= 1.691\text{x}10^5 + 21986\,Q - 255.14\,Q^2 + 1.5825\,Q^3 \quad \text{for } n=10,\ Q > 6 \text{ gpm} \tag{4.11}$$

$$= 1.375\text{x}10^5 + 16586\,Q - 183.79\,Q^2 + 1.144\,Q^3 \quad \text{for } n=20,\ Q > 6 \text{ gpm} \tag{4.12}$$

The capital cost of WAO is in the millions, but the payback may occur within a year for treatment of large volumes of waste. When it is compared to incineration, the capital cost for WAO is greater, but the operating costs are lower and there has been little public resistance to the installation of WAO, while public resistance to the installation of incinerators is often intense. In addition, for some types of wastes, the components leaving the reactor may be saleable, as is the case with the production of ammonium sulfate in the steel industry.

4.4.3. Results of Analysis of the Suitability of Wet Air Oxidation

The results of the treatability and cost estimation analysis are summarized in Table 4.9, which contains data on wastestreams treatable by wet-air oxidation. As mentioned above, each treatable stream has been classified as A,B, or C according to the influent concentration of treatable contaminants with A corresponding to 0-1%, B to 1-3% and C to 3-5%. These categories likely represent decreasing overall efficiencies of contaminant removal. Nevertheless, some streams categorized as C may display very high removal efficiencies while some categorized as A may have low removal efficiencies.

Of the 157 California wastestreams treatable by wet-air oxidation (Table 4.9), 106 were classified as A, 32 forms as B and 18 forms as C, indicating that most of the wastestreams (68%) in Table 4.9 can be treated with high removal efficiencies. About 37% of the treatable wastestreams are generated by the metal-finishing industry with SIC codes 34 and 36. This result is reasonable, since the wastes from such industries often contain cyanides at levels that can be effectively removed by wet-air oxidation. The remaining forms represent wastes containing organic compounds or solvents which can also be treated with WAO (e.g., SIC codes 28, 29 and 37). Table 4.9 also indicates whether catalyst is needed for oxidation of the wastestream.

The capital costs shown in Table 4.9 are the minimum capital costs (except two), indicating that all the treatable wastestreams had small flow rates (< 6 gal/min.). Similarly, the annual operating costs for 10 and 20 years all fell between $ 210,000 and $ 280,000, with the exception of the two larger wastestreams mentioned above.

As mentioned previously, due to the simplicity of the selection criteria for wet air oxidation in this study, the results of the selection and costs should be viewed with great caution.

4.5. EMERGING TECHNOLOGY IN WET AIR OXIDATION -- SUPERCRITICAL WATER OXIDATION

Recently, it has been discovered that if the usual operating temperature (175-325° C) and pressures (500-2000 psig) for wet air oxidation were raised above the critical condition of water (supercritical condition), the oxidation potential of the process, and consequently the removal efficiency, would increase greatly.

As a medium for chemical reaction, supercritical water (SCW) has been found very useful for oxidation of organics. Above the critical temperature (374 ^{o}C) and pressure (3230 psig), water is an excellent solvent for both organic compounds and gases (e.g. oxygen, nitrogen, carbon dioxide) [12,13]. Dissolving waste in air or oxygen in SCW leads to a homogeneous phase in which oxidation is rapid. At 400° C to 450° C, 99 to 99.9% conversion (to carbon dioxide, carbon monoxide) can be obtained within 5 minutes. If the operating temperatures are raised to 600-650° C, 99.9999% can be achieved in less than a minute [10]. Table 4.10 shows the removal efficiencies of some selected organics oxidized by supercritical water [10].

Oxidyne [11] has proposed to conduct wet air oxidation and supercritical water oxidation in reactors which are placed underground in deep, well-like cavities. The process has been referred to as "downhole" oxidation. A schematic diagram of the reactor is shown in Figure 4.10. In the reactor, oxygen is injected into the liquid organic waste at sufficient pressure and temperature to support liquid state oxidation. This continuous flow system converts the organic waste into inert ash, carbon dioxide and water. In addition, the excess heat generated may be recovered at the surface for increased energy efficiency. The Oxidyne process, when used to treat municipal sludges, can achieve 95% COD reduction, 93% BOD (biological oxygen demand) reduction and 98% reduction of volatile organics [11]. The process is very efficient for treating wastes with less than 10% organics. The estimated cost for treatment varies with flow rates, ranging from $ 0.27/gal for 3 gpm to $ 0.11/gal for 174 gpm [11].

Table 4.10. Removal Efficiencies of Organics by Supercritical Water Oxidation

Class/ Compound	Temperature, °C	Residence Time, Min.	Removal Efficiency, %
Aliphatic Hydrocarbons			
Cyclohexane	445	7.0	99.97
Aromatic Hydrocarbons			
Biphenyl	450	7.0	99.97
o-Xylene	495	3.6	99.93
Halogenated Aliphatics			
1,1,1-Trichloroethane	495	3.6	99.99
1,2-Ethylene dichloride	495	3.6	99.99
1,1,2,2-Tetrachloroethylene	495	3.6	99.99
Halogenated Aromatics			
o-Chlorotoluene	495	3.6	99.99
Hexachlorocyclopentadiene	488	3.5	99.99
1,2,4-Trichlorobenzene	495	3.6	99.99
4,4-Dichlorobiphenyl	500	4.4	99.993
DDT	505	3.7	99.997
PCB 1234	510	3.7	99.99
PCB 1254	510	3.7	99.99
Oxygenated Compounds			
Methyl ethyl ketone	460	3.2	99.96
Methyl ethyl ketone	505	3.7	99.993
Dextrose	440	7.0	99.6
Organic Nitrogen Compounds			
2,4-Dinitrotoluene	457	0.5	99.7
2,4-Dinitrotoluene	513	0.5	99.992
2,4-Dinitrotoluene	574	0.5	99.9998

Source: [13]

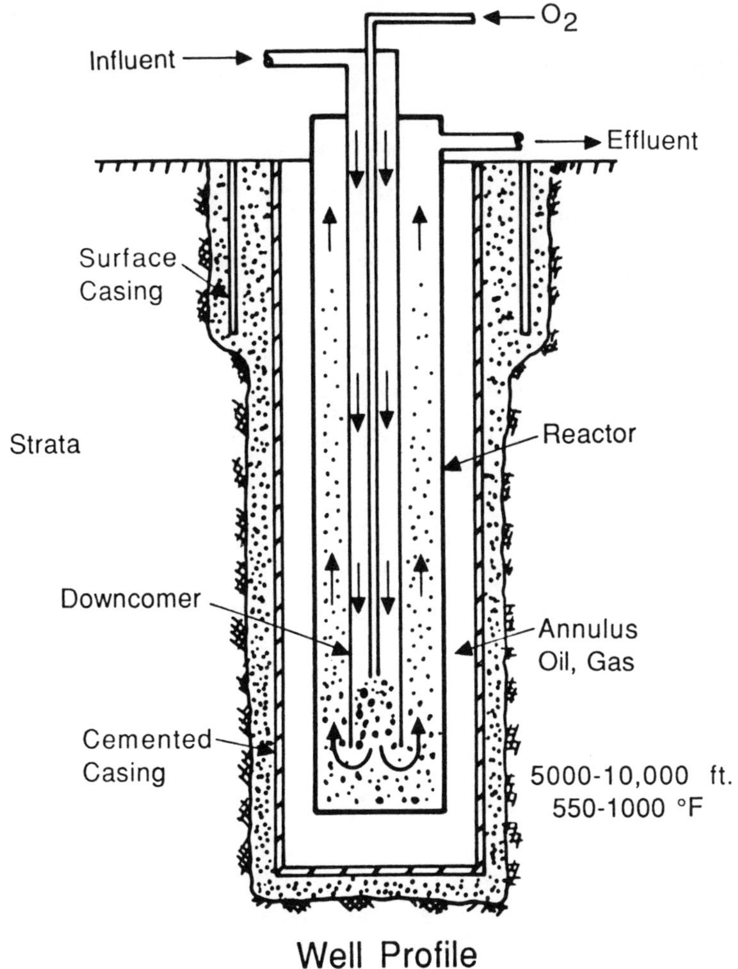

Figure 4.10 Schematic Diagram of "downhole" Oxidation Reactor.

4.6. REFERENCES

1. Heimbuch, J., Wilhelmi, A.R., "Wet Air Oxidation - A Treatment Means for Aqueous Hazardous Waste Streams", J. of Hazardous Materials, December, 1985, p. 187.

2. Balog, S.E., Kerr, R.K., Pradt, L.A., "The Wet Air Oxidation Boiler for Enhanced Oil Recovery", Canadian Petroleum Technology, Sept-Oct 1982, pp.73.

3. Wilhelmi, A.R., Copa, Ellis, "Wet Air Oxidation of Hazardous Wastes", Technical Bulletin from Zimpro Inc., May 1987.

4 Randall, T.R., "Wet Oxidation of Toxic and Hazardous Compounds", Zimpro Inc. Technical Bulletin, I-610, June 29-30, 1981.

5. Wilhelmi, A.R., Knopp, P.V., "Wet Air Oxidation - An Alternative to Incineration", Chemical Engineering Progress, August 1979.

6. Oettinger, T.P., Hoffman, M.C., Fontana, M.G., "Use of Titanium for Treatment of Toxic and Hazardous Wastes", Industrial Applications of Titanium and Zirconium," vol. 4, American Society for Testing and Materials, 1986, p. 28.

7. "Helping Steel Kick The Coke Oven Gas Habit", The Reactor Magazine, no. 41, February 1979.

8. Bell, R.L., Jackman, A.P., and Powell, R.L.,a report submitted to the California Dept. of Health Services by Dept. of Chemical Engr., U. of California, Davis, 1987.

9. McHugh, M. A., Krukonis, V.J., Supercritical Fluid Extraction, Butterworths Publishers, Stoneham, MA., 1986.

10. Modell, M., "Supercritical Fluid Technology in Hazardous Waste Treatment", Conference Proceedings of Oil Waste Management Alternatives Symposium in Oakland, CA, April 28-29, 1988.

11. Technical Bulletin on Deep-well Oxidation from Oxidyne Co.

12. "Positive Results with Wet Air Oxidation", The Reactor Magazine, no. 41, February 1979.

13. Dietrich, M.J., Randall, T.L., Canney, P.J., "Wet Air Oxidation of Hazardous Organics in Wastewater", Environmental Progress, vol. 4, n. 3. August, 1985.

14. DeClerck, D.H., Patarcity, A.J., "32nd Biennial Report on Materials of Construction", Chemical Engineering, Nov. 24, 1986.

5. Chemical Fixation and Solidification

5.1. INTRODUCTION

The use of chemical fixation in the treatment of hazardous wastes offers several advantages, such as improvement in the handling characteristics of the waste, by producing a solidified material which has sufficient strength for land disposal [1]. Ideally, the modified structure of the final product significantly reduces the leachability of the toxic contaminants. If leachability is reduced to the level which satisfies the Environmental Protection Agency's Extraction Procedure (EP) Toxicity Test, or, in California, the Wet Extraction Test (WET), the "fixed" or "solidified" waste is considered to be nonhazardous [2].

Chemical fixation, for the most part, has become synonymous with stabilization. However, there is a fine distinction between chemical fixation and stabilization. **Chemical fixation** is the chemical technology used to detoxify, immobilize, insolubilize or otherwise render a waste component less hazardous when introduced into the environment. It often denotes a chemical reaction between one or more waste components and a solid matrix that is either deliberately introduced or is already in the waste residue [3].

Solidification refers to a pretreatment process that produces a solid monolithic mass with improved structural integrity and physical characteristics that are more suitable for transportation, storage, landfill or reuse [2,3]. Solidification can occur without chemical fixation, and it does not, therefore, affect the toxicity of the waste. It may, however, reduce the hazard by setting up a barrier between the waste particles and the environment, such that the permeability of the contaminants by water is reduced, in effect, reducing the potential for mass transfer [3]. In this Chapter, we use "chemical fixation" and "stabilization" interchangeably.

Almost all chemical fixation processes involve reactions between the waste and some proprietary additives that promote precipitation of soluble metal ions as insoluble hydroxides [4]. These additives can enhance the curing reaction that solidifies the waste and it can increase binding between the waste and the solidifying reagents. The details of a stabilization (fixation) process vary depending upon the nature of the waste being treated. In turn, the physical and chemical properties of the treated products vary according to: (a) characteristics of the waste, (b) types of additives and solidifying reagents, (c) drying conditions, and (d) curing time [4]. Most existing solidification/stabilization technologies were developed to treat inorganic wastes, primarily those that contain dissolved metals. Organic compounds generally interfere with the setting and curing of the solidifying agent [5]. However, as will be seen in the vendors' surveys, certain chemical fixation processes are claimed capable of treating wastes that contain up to 80% organic compounds by using proprietary additives.

The primary goals of hazardous waste fixation include: (a) improved handling and physical characteristics, (b) decreased surface area across which the transfer of contaminated pollutants can occur, and (c) limited solubility of the pollutants contained in the waste [6]. These goals can be achieved in different ways, but not all techniques can meet all three goals. A particular fixation technique may adequately treat a particular type of waste but may not be suitable for others. The present fixation processes can be grouped into the following seven categories: [6]

(a) Cement-based

(b) Lime-based

(c) Thermoplastic (including bitumen, paraffin and polyethylene)

(d) Reactive polymer

(e) Encapsulation

(f) Self-cementing

(g) Glassification

The important attributes, advantages and disadvantages of these seven categories are discussed in the subsequent sections. Since fixation systems vary widely in the applicability, cost and pretreatment requirements, many are limited as to the types of waste that can be economically processed. Selection of any particular technique for waste fixation must include careful consideration of the containment required, the cost of processing, the increase in bulk volume of the material and the changes in the handling characteristics. Among the seven processes, the cement-based and lime-based techniques are the most popular and least expensive treatments. A few commercial processes utilizing these will be discussed in Section 5.4.

5.2. DESCRIPTION OF CHEMICAL FIXATION PROCESSES

5.2.1. Cement-Based Techniques

Cement-based fixation techniques generally use Portland cement. This cement is a fused mixture of calcium, silicon, aluminum and iron oxides, the main constituents of which are usually tri- and di-calcium silicates [7]. The sludge or slurry to be treated, along with other additives (sometimes proprietary) including fly ash or other aggregate, is mixed with the cement to form a monolithic, rock-like mass. The wet waste and sludges initially form a gel. As the solidification reaction proceeds, fibrils are formed upon hydration of the silicate compound. These interlocking fibrils bind the cement and various hydration products into the hardening mass. The success of the hardening process is significantly influenced by the presence of compounds such as sulfates, borates, salts and a variety of organics [6]. The end product may be a standing monolithic solid or may be a crumbly, soil-like material, depending upon the physical and chemical properties of the waste and the amount of cement used.

Five different types of Portland cement are generally used. These are differentiated on the basis of variations in the chemical compositions and the physical properties: [8]

(a) Type I is the "normal" cement used for construction purposes and it constitutes over 90% of the cement manufactured in the United States.

(b) Type II is used in the presence of moderate sulfate concentrations (150-1500 mg/kg).

(c) Type III has a high strength cement and is used when rapid setting is required.

(d) Type IV produces a low heat of hydration and is used in large concrete work.

(e) Type V is a special low alumina, sulfate-resistant cement used for high sulfate concentrations (>1500 mg/kg).

Most waste fixation is done with Type I Portland cement, but Types II and V are often used for sulfate or sulfite wastes.

Most hazardous wastes which are aqueous slurries can be mixed directly with cement and the suspended solids will be incorporated into the rigid matrix. Cement solidification is most suitable for treating inorganic wastes, especially those containing metals. When the cement mixture has a high pH, most multivalent cations are converted into insoluble hydroxides or carbonates [4]. Certain materials in the waste, such as latex and solid plastics, may act as reinforcing agents and increase the strength and stability of the waste concrete while other materials like silt, clay, coal or lignite may delay setting and curing of the cement [7]. Other compounds which are especially active as retarders of the setting process include sodium salts of arsenate, borate, phosphate, iodate and sulfide. Products containing large amounts of sulfate not only retard the process but also react with it to form sulfoaluminate hydrate, which causes swelling and spalling in the solidified waste concrete. Under such circumstances, the special low alumina (Type V) cement is needed to inhibit this reaction.

A number of additives have been developed for use with cement to improve the physical characteristics, for example, strength, and also to decrease the leaching losses from the solidified mass. Many of the additives are proprietary. Some fixation processes that use specialized additives will be discussed in Section 5.4. Very often these additives contain clay or vermiculite, which are used to absorb excessive free liquid, or they contain sodium silicate to increase the binding (through chemical reactions with the cement) between the waste and the cement. Unfortunately, the additives can cause an increase in the final volume of the setting mixture and, hence, increase the cost of ultimate disposal [6]. Table 5.1 summarizes the compatibility of the selected waste categories with the cement-based process.

Some of the advantages of cement-based fixation/solidification systems are: [6]

(a) Raw materials (cement) are plentiful and inexpensive.

(b) Cement mixing and handling are well developed; and the processing equipment is readily available. Special labor is not required.

Table 5.1. Applicability of Cement-based Processes to Selected Waste Categories

Waste Component	Degree of Applicability
Organics	
Organic solvents and oils	Many impede setting, may escape as vapor.
Solid organics (e.g. plastics, resin, tars)	Good - often increase durability.
Inorganics	
Acid wastes	Cement will neutralize acids.
Oxidizers	Compatible.
Sulfates	May retard setting and cause spalling unless special cement is used.
Halides	Easily leached from cement, may retard setting.
Heavy metals	Compatible.
Radioactive materials	Compatible.

Source: [1]

(c) Extensive drying or dewatering of waste is not needed. Cement takes up water during mixing, and the amount of cement can be varied to adapt to a wide range of water content.

(d) Cement is reasonably tolerant of chemical variations. The natural alkalinity of the cement can neutralize acids in waste. In addition, cement is not affected by strong oxidizing agents like nitrates or chlorates. Pretreatment is necessary only for materials that retard the setting of the cement.

Some of the disadvantages are:

(a) Relatively large amounts of cement are required for most fixation processes. The final weight and volume of the product may be double those of other fixation techniques, such as thermoplastic and organic polymer processes.

(b) Low-strength cement waste mixtures are often vulnerable to leaching by acidic solutions. Extreme conditions can result in decomposition of the product and accelerated leaching of the contaminants.

(c) Pretreatment, expensive cement types, or costly additives may be necessary for stabilization of wastes that contain impurities which can affect the curing and setting of the cement.

5.2.2. Lime-Based Techniques

Waste fixation techniques using lime-products usually depend on the reaction of lime with a finely-graded siliceous (pozzolanic) material and water to produce a concrete-like material (sometimes referred to as a *pozzolanic concrete*). Hydrates of calcium silicates, calcium aluminates, or calcium aluminosilicates are formed by the reaction of calcium in lime with amorphous aluminosilicates. The pozzolanic products formed entrap the wastes. In seven to twenty-eight days a nearly impermeable mass is developed, similar to the processes using cement [7].

The most common pozzolanic materials are fly ash, blast furnace slag, ground brick, and cement-kiln dust [4]. All of these materials are themselves waste products with little or no value. Therefore, the use of these waste products to consolidate with another waste is often an advantage to the generator, who can dispose two wastestreams at the same time.

The types of additives that are usually used for lime-based chemical fixation include: [7]

(a) Certain clays which absorb liquid and bind specific anions or cations.

(b) Emulsifiers and surfactants which allow the incorporation of immiscible organic liquids.

(c) Proprietary absorbents like carbon, zeolite materials or cellulosic sorbents that selectively bind specific wastes.

Table 5.2 summarizes the feasibility of using the lime-based technique to treat the same types of wastes described in Table 5.1. Only organics and halides cause problems in the setting of the mixture. However, with the introduction of specialized additives, such problems can be overcome.

Some of the advantages with this process are:

(a) Materials are generally inexpensive and widely available.

(b) Processing equipment is simple to operate and readily available.

(c) Excessive dewatering is not necessary since water is required in the setting.

(d) Chemistry of lime-pozzolanic reactions is well-known.

The lime-based systems also have many of the same disadvantages as cement-based processes:

(a) Lime and other additives add to the weight and bulk of waste to be transported and/or landfilled.

(b) Uncoated lime-fixed materials may require specially designed landfills to ensure that the material does not lose potential pollutants by leaching.

Table 5.2. Applicability of Lime-Based Processes to Selected Waste Categories

Waste Component	Degree of Applicability
Organics	
Organic solvents and oils	Many impede setting, may escape as vapor.
Solid organics (e.g. plastics, resin, tars)	Good - often increases durability.
Inorganics	
Acid wastes	Compatible.
Oxidizers	Compatible.
Sulfates	Compatible.
Halides	May retard setting, most are easily leached.
Heavy metals	Compatible.
Radioactive materials	Compatible.

Source: [1]

5.2.3. Thermoplastic Processes

In processes using thermoplastics dried wastes are mixed with a molten thermoplastic material such as bitumen, asphalt, paraffin, polyethylene, or polypropylene. Asphalt is the most commonly used material and it is suitable for heavy metal or electroplating wastes. Depending upon the particular type of thermoplastic material used, the operating temperature of this process usually ranges from 130 to 230°C [4]. Once the blended mixture is cooled, the thermoplastic medium hardens, thereby producing a matrix that adheres well to the incorporated waste. The resulting waste has leach loss rates which are significantly lower than other fixation techniques, and it is quite resistant to aqueous solutions and microbial degradation [7].

The applicability of thermoplastic encapsulation is limited by the presence of organic chemicals, which act as solvents towards the binder, and strong oxidizers (nitrates, chlorates or perchlorates), which react with the organic matrix and cause slow deterioration [6]. The technique is also not suitable for any anhydrous compound that can rehydrate easily after treatment, causing the thermoplastic matrix to split apart and thereby increasing the surface area and the rate of waste loss [6]. Table 5.3 lists the applicability of solidification using thermoplastics for treating the wastes considered in Tables 5.1 and 5.2.

Some of the major advantages of using a thermoplastic matrix are:

(a) Leaching rates of the contaminants from the treated mixture are significantly lower than those from the cement-based or lime-based processes.

(b) Overall volume of the waste may be reduced, since the waste needs to be dewatered before using the thermoplastic technique.

(c) Thermoplastics usually adhere well to the materials being encapsulated.

(d) End-product is fairly resistant to attack by aqueous solutions. Microbial degradation is minimal.

(e) Materials embedded in the thermoplastic matrix can be reclaimed if needed.

(f) End-product will tend to be lighter than if a cement-based system is used since the weight of the thermoplastic matrix is less. This low density would reduce the

Table 5.3. Applicability of Thermoplastic Processes to Selected Waste Categories

Waste Component	Degree of Applicability
Organics	
Organic solvents and oils	Organics may vaporize on heating.
Solid organics (e.g. plastics, resins, tars)	Possible use as binding agents.
Inorganics	
Acid wastes	Can be neutralized before incorporation.
Oxidizers	May cause matrix breakdown, fire.
Sulfates	May hydrate or rehydrate, causing splitting.
Halides	May dehydrate.
Heavy metals	Compatible.
Radioactive materials	Compatible.

Source: [1]

transportation costs on a per weight of treated stream basis.

The disadvantages of using a thermoplastic matrix are:

(a) Expensive equipment and skilled labor are necessary for processing.

(b) They cannot be used with materials that decompose at high temperatures, especially citrates and certain types of plastics.

(c) Thermoplastic materials are flammable. There are workplace hazards associated with working with organic materials such as bitumen at elevated temperatures.

(d) During heating, some mixtures that contain volatile organics can produce objectionable oils and odors, causing secondary air pollution.

(e) The waste materials must be dried before they can be mixed with the thermoplastic materials. This requires a large amount of energy. Incorporating wet wastes greatly increases losses through leaching and poses the possibility of considerable gas (steam) evolution during processing. This steam likely carries with it dissolved contaminants which pose both workplace and general air quality hazards.

(f) Strong oxidizers are usually not suitable for thermoplastic processes due to the possible reaction between the oxidizers and the binding material.

(g) Dehydrated salts within the thermoplastic matrix slowly rehydrate if the mixture is soaked in water. This causes the waste block to fragment, thus increasing the rate of leaching.

5.2.4. Reactive Polymer Processes

In contrast with the thermoplastic techniques in which a polymerized material is heated and mixed with the substance to be solidified, reactive polymer processes usually are carried out at ambient temperature. They involve the mixing of monomers, such as urea-formaldehyde, with a catalyst to form a polymer. The polymer is formed in a batch reactor where the wet or dry waste is blended with a prepolymer using a specially designed mixer [6]. When the two components are thoroughly mixed, a catalyst is added and mixing is continued until the catalyst is completely dispersed. Mixing is terminated before the polymer forms and the resin-waste mixture is transferred to a waste container, if necessary. The polymerized material does not chemically combine with the waste; it forms a sponge-like mass that traps solid particles but leaves liquid wastes alone. The polymer mass is usually dried in order to increase the binding between the polymer and the waste prior to land disposal [7]. Reactive polymer processes generally require less solidifying agent per weight of waste than do other solidification systems, and they produce a less dense material for disposal [4]. The degree of binding (cross-linking) between the waste and the polymer is influenced by parameters such as pH, water content, and ionic constituents in the feed stream [1].

Several alternative polymers have been used for the reactive polymer technique including: ureas, phenolics, epoxides, polyesters and vinyls [7]. Table 5.4 gives a summary of the applicability of the organic polymer technique in treating those wastes considered in Tables 5.1-5.3.

Table 5.4. Applicability of Reactive Polymer Processes to Selected Waste Categories

Waste Component	Degree of Applicability
Organics	
Organic solvents and oils	May retard setting of polymers.
Solid organics (e.g. plastics, resins, tars)	May retard setting of polymers.
Inorganics	
Acid wastes	Compatible.
Oxidizers	May cause matrix breakdown.
Sulfates	Compatible.
Halides	Compatible.
Heavy metals	Acid solutions may dissolve metal hydroxides.
Radioactive materials	Compatible.

Source: [1]

The major advantages of the reactive polymer process are:

(a) Less fixative material is required for solidifying the same amount of waste than using cement or lime-based techniques.

(b) Waste material is usually dewatered, but not necessarily completely dried. The finished, solidified polymer, however, must be dried before ultimate disposal, with the resulting reduction in the amount of waste to be disposed.

(c) End-product has a lower density (specific gravity ~1.3) than cement. The low density reduces the transportation costs for the fixed product.

(d) Solidified resin is non-flammable and high temperature is not required to form the resin.

The major disadvantages of the reactive polymer process include:

(a) No chemical reactions occur in the solidification process that chemically bind the potential pollutants. The particles of the waste material are simply entrapped in an organic matrix.

(b) Catalysts used in the urea-formaldehyde process are strongly acidic. Most metals are fairly soluble at low pH and can escape in water not trapped in the mass during the polymerization process. The catalyst may be highly corrosive and require special mixing equipment, reactors, containers, etc.

(c) Some of the reactive polymers are biodegradable.

(d) Secondary containment in steel drums is required before disposal, raising costs in processing and transportation.

5.2.5. Encapsulation Technique

Once the wastes have been solidified by drying, a second level of containment can be imposed by coating the solidified waste with a layer of protective, impermeable material. The encapsulating material is usually an organic polymer such as polyethylene, polybutadiene, polyurethane, or fiberglass/epoxide [7]. The goal is to create an encapsulated waste which is a strong solid, resistant to chemical and mechanical attack and to water penetration. This enables highly soluble wastes to be isolated from the environment, thereby greatly reducing leaching. On the other hand, care must be taken to ensure adhesion between the coating and the waste and to use a coating material which, will have long-term integrity in encapsulating the waste [6].

The applicability of such an encapsulation process is listed in Table 5.5. When ideally implemented, its major advantage is that the waste materials never come in contact with water and very soluble materials can be successfully isolated from the environment. A secondary container (steel drum, etc) is usually not required because the coating materials are strong and chemically inert.

The major disadvantages of this technique are:

(a) Materials used for encapsulation are expensive.

(b) Large quantities of energy for drying, fusing the binder, and forming the jacket are required.

(c) Certain jacket materials are combustible.

(d) Extensive capital investments in specialized equipment are required.

(e) Skilled labor is required to operate molding and fusing equipment.

Table 5.5. Applicability of Encapsulation Processes to Selected Waste Categories

Waste Component	Degree of Applicability
Organics	
Organic solvents and oils	Must first be adsorbed on solid matrix.
Solid organics (e.g. plastics, resins, tars)	Compatible - many encapsulating materials are plastics.
Inorganics	
Acid wastes	Can be neutralized before incorporation.
Oxidizers	May cause deterioration of encapsulating materials.
Sulfates	Compatible.
Halides	Compatible.
Heavy metals	Compatible.
Radioactive materials	Compatible.

Source: [1]

5.2.6. Self-Cementing Processes

Some industrial wastes, such as flue gas cleaning or desulfurization sludges, that contain high concentrations of calcium sulfate or calcium sulfite can be treated using a *self-cementing* process [4]. A small portion of the dewatered waste is heated (calcined) under carefully controlled conditions to produce a partially dehydrated cementitious calcium sulfate or sulfite which is used as the binding material. This calcined waste is reintroduced into the waste sludge along with proprietary additives. Fly ash is added to adjust the moisture content. The finished product is a hard, plaster-like material with good handling characteristics and low permeability [6].

The major advantages of this process are:

(a) No major additives have to be manufactured and shipped to the processing site.

(b) Product setting time and curing times are shorter than those required in the comparable cement-based and lime-based systems.

(c) Heavy metals can be effectively retained.

(d) These systems do not require completely dry waste. The hydration reaction uses water.

The major disadvantages of the self-cementing technique are:

(a) Only high sulfate or sulfite sludges can be treated.

(b) Self-cemented sludges have much the same leaching characteristics as the cement-based and lime-based processes.

(c) Extra energy is required to produce the calcined cementitious material.

(d) The process requires skilled labor and special equipment to calcine the waste and mix it with additives.

5.2.7. Glassification

Wastes that are stable at very high temperatures can be fused into glass or ceramics. This process is most popular for the stabilization of radioactive wastes and cannot be used to treat organic wastes. Since glasses are not leachable by water, this process is generally assumed to produce a safe material for disposal without secondary containment [4].

Some advantages of glassification are that the process produces a high degree of waste containment, and that the additives (syenite and lime) which are used are relatively inexpensive.

The major disadvantages of the process are:

(a) Process is energy-intensive. The glass must be heated to 1350°C to obtain the molten form required to fuse with the waste.

(b) Some constituents, especially metals, may vaporize due to the high temperature before they can combine with the molten silica in the glass.

(c) Specialized equipment and trained personnel are required.

5.3. EVALUATION OF STABILIZED/SOLIDIFIED WASTE PRODUCTS

The success of a stabilization process in treating a waste is evaluated by testing the stabilized/solidified products. The physical and chemical properties of the stabilized wastes are typically used to indicate the level of containment of waste chemicals, and to various degrees, to predict durability in actual field conditions [7]. Moreover, both types of properties significantly influence the ultimate strategy for final disposition of the waste, which can represent a substantial percentage of the overall cost of treatment. A discussion of physical and chemical properties of the stabilized/solidified waste is presented in this section.

5.3.1. Physical Tests of Stabilized Wastes

Physical tests of the solidified waste are aimed at: (a) determining the particle size distribution, porosity, permeability, and wet and dry densities; (b) predicting the response of the material to stresses which might be applied in embankments, landfills, etc.; and (c) evaluating durability [7]. Most of these tests are described in the Annual Book of American Society for Testing and Materials (ASTM) Standards and the Engineering Manual [9,10].

Five standardized tests are frequently applied to stabilized products. These include bulk and dry unit weight, unconfined compressive strength, permeability, wet/dry durability, and freeze/thaw durability [7]. Bulk and dry unit weight are indirect measures of density and void volume. The bulk unit weight is obtained by measuring the weight per unit of total volume (solids and water), while the dry unit weight is determined from the division of the oven-dried weight by the total volume. The unconfined compressive strength is the maximum axial compressive stress at failure or at 15% strain, depending on which occurs first. The permeability of the treated waste is an indication of the ability of a material to allow or permit the passage of water. Since permeability is the ability of a material to conduct or discharge water when placed in a hydraulic gradient, it is a measure of the likelihood of releasing contaminants to the environment. The wet/dry durability test evaluates the resistance of the material to natural weathering stresses of wetting and drying. During the test, twelve wet-dry cycles are applied, and weight loss measurements are taken to determine the amount of sample degradation. Freezing and thawing are two other natural weathering stresses which are important in evaluating the structural integrity of the stabilized waste. The sample is subjected to twelve freezing/thawing cycles with each followed by measurements of the weight.

Table 5.6 lists the the sources of the test protocols for each of the five physical parameter tests. A summary of the results of these tests on a variety of wastes using different fixation techniques has been tabulated and is given in Table 5.7 [6]. In nearly all cases, all of the fixation processes increased density and strength while decreasing permeability. Generally, cement- and lime-based products resulted in high compressive strength and low permeability, while plastic encapsulation resulted in a material having relatively low permeability.

Table 5.6. Selected Tests of Physical Properties

Test	Source
Bulk and dry unit weight	Appendix II of EM-1110-2-1906
Unconfined compressive strength	Appendix XI of EM-1110-2-1906 and ASTM Method D2166-66
Permeability	Appendix VII of EM 1110-2-1906
Wet/dry durability	ASTM Method D599-57
Freeze/thaw durability	ASTM Method D560-57

Source: [9,10]

Table 5.7. Results for Physical Testing of Stabilized and Untreated Industrial Wastes

Type of waste and treatment	Unit weight Bulk (lb/ft^3)	Unit weight Dry (lb/ft^3)	Unconfined compressive strength (lb/in^2)	Permeability (cm/s)	Durability (test cycles to failure) Wet/dry	Durability (test cycles to failure) Freeze/thaw
Nickel-cadmium battery waste						
Untreated	--[a]	43.9	–	5.7×10^{-6}	–	–
Lime-based pozzolan product	104.0	86.2	169.0	1.9×10^{-6}	9	–
Patented additives, soil-like material	93.2	47.3	7.96	1.9×10^{-4}	1	1
Chlorine production waste						
Untreated	--	64.0	–	1.0×10^{-4}	–	–
Lime-based pozzolan product	103.1	88.6	133.0	8.5×10^{-7}	–	–
Patented additives, soil-like material	106.0	81.3	21.6	3.6×10^{-5}	2	1
Calcium fluoride waste						
Untreated	–	46.8	–	3.5×10^{-5}	–	–
Lime-based pozzolan product	85.9	66.6	26.2	3.8×10^{-5}	1	1
Patented additives, soil-like material	86.2	52.8	25.6	8.7×10^{-6}	1	2
Electroplating waste						
Untreated	–	28.1	–	3.1×10^{-5}	–	–
Lime-based pozzolan product	100.0	77.4	77.3	4.0×10^{-7}	5	–
Patented additives, soil-like material	87.1	47.4	32.4	1.1×10^{-5}	2	–
Organic resin, rubber-like material	75.4	52.7	747.0	1.1×10^{-4}[b]	1	12
Plastic encapsulation	73.6	73.6	1540.0	impervious	NF(0.00)[c]	NF(0.00)[c]
Flue gas cleaning waste						
Untreated	–	58.8	–	3.6×10^{-5}	–	–
Lime-based pozzolan product	100.1	80.9	100.0	2.0×10^{-6}	3	2
Patented additives, soil-like material	77.0	43.4	23.7	1.6×10^{-4}	1	1
Cemented-based, concrete-like product	101.0	94.9	2570.0	7.9×10^{-4}[b]	NF(15.8)[c]	10

[a] --, not tested.
[b] Value questionable because flow restriction caused by sample support may have influenced flow through sample.
[c] NF indicates no failure in 12 cycles; figures in parentheses are the percent weight loss after 12 cycles.
Source: [7]

5.3.2. Chemical Tests of Stabilized Waste

The leaching test is the primary and most widely used indicator of chemical stability and thereby the potential environmental effects of the treated waste. Leachability is measured by exposing a waste, treated or not, to a given solution and determining its rate of dissolution. From the standpoint discussed here, leaching is the rate at which hazardous components are removed from the waste and contaminate the leachant [3]. (The solution that does the leaching is called the leachant, and the contaminated solution that passed through the waste is the leachate.)

Measurements of the leaching rate are usually reported in terms of concentration of the constituent in the leachate. Contaminant concentration, which is the primary regulatory basis for water-quality standards, is also used as the basis for leaching standards. When evaluating the degree of leachability of a material, a comparison is made between the concentration of the hazardous constituent in the leachate and that in the original waste [3]. This indicates the proportion of the contaminant that was released during the test. More information can be obtained if the time for the leaching test is known. Then, the true rate of leachability of the waste constituents can be determined. The primary variables that affect the outcome of the leaching tests are temperature, nature of leaching solution, waste-to-leaching solution ratios, number of extractions, time of extraction, surface area of waste, and agitation technique [7]. With the exception of the number of extractions, when values of other parameters are raised (e.g., temperature, acidity,waste-to-leaching solution ratio, etc.), the concentration of contaminant in the leachate almost invariably increases.

To have a waste designated nonhazardous, or delisted, the waste generator must demonstrate that the hazardous constituents cannot be leached from the treated product at concentrations which would still be considered hazardous. In the United States, this generally means that any hazardous constituent cannot be present in the leachate at a concentration greater than 100 times the drinking water standards [3].

Presently, a number of leachate tests are used with no single standardized test adequately predicting the potential impact of a stabilized waste on the environment. The choice of an applicable leaching test is determined by the company or agency handling the waste and/or testing the fixation process. The conditions under which extraction tests are conducted are intended to simulate the environments to which wastes might be exposed after disposal in a landfill area. The most widely used leaching tests for solidified and other waste materials include the Environmental Protection Agency's EP (Extraction Procedure) Toxicity Test [11], Toxic Characteristic Leaching

Procedure [12] (TCLP, a modified EP Toxicity Test), and Multiple Extraction Procedure [13] (MEP). At the state level, a typical test is the Wet Extraction Test (WET) introduced by the California Department of Health Services [4].

The EPA EP Toxicity Test [11] procedure uses 0.5N acetic acid to enhance leaching of hazardous constituents. The solidified end product is first pulverized and passed through a No. 4 (4.76 mm) U.S. standard sieve. The rate of leaching of contaminants from the stabilized waste, is determined by placing 200 g of the pulverized sample in a large plastic bottle with 2 liters of distilled deionized water. The pH of the slurry is then lowered and maintained at 5 by the addition of 0.5N acetic acid. The test proceeds for twenty four hours during which time the slurry is kept well-mixed using a rotary shaker. The cumulative amount of acid added cannot exceed 4 ml per gram of sample or 800 ml for the 200 g sample tested. The sample is removed from the shaker after twenty four hours and filtered through a 0.45 micron filter. The leachate is then analyzed for whatever contaminants that were known to be in the original sample. Table 5.8 gives the results of the EP leaching test on stabilized metal hydroxide sludges treated with cement. Both metals in the leachate, cadmium and lead, were present at concentrations that were below the EP Toxicity limits (1 mg/l and 5 mg/l, respectively).

The Toxic Characteristic Leaching Procedure [12], (TCLP) is similar to the EP Toxicity Test, with the major difference being that in the TCLP, organic compounds are also leached. The minor differences of TCLP from EP Test are the elimination of continual pH adjustment and shortening the duration of test in TCLP.

The Multiple Extraction Procedure [13] provides a tougher and more realistic test than the EP Procedure. This procedure includes a series of contaminant extractions after the EP test to check the long-term integrity of the solidified waste after disposal. An acid solution is prepared by adding a 60/40 weight percent mixture of sulfuric and nitric acid to distilled water until a pH of 3 is achieved. The waste sample after the EP leaching test is weighed and placed in the shaker with twenty times its weight of the extraction fluid. The mixture and sample are then agitated for twenty four hours at a constant temperature between 68 and 104°F. The pH is recorded five to ten minutes prior to extraction and at the end of extraction. At the end of the twenty-four hour extraction period, the contents are separated into the liquid and solid phases in the EP leaching test. The extract is then analyzed for barium, cadmium, chromium, lead, mercury, nickel, silver, arsenic, selenium, cyanide and phenol. The remaining solid phase then undergoes eight subsequent extractions in accordance with the above steps. This Procedure is one of the most rigorous and, in some circumstances, it may be more severe than natural processes. Table 5.9 summarizes the results of a multiple extraction test on a cement-fixed metal sludge [15]. Again, the concentrations of the heavy metals in each extraction cycle meet the EP Toxicity limits.

Table 5.8. Metal Concentrations in EPA-EP Test of Waste Metal Hydroxide Sludge Mixed with Cement

Metal	Sludge/Cement ratio	Metal Conc. in Sludge, mg/l	Metal Conc. in Leachate, mg/l
Cadmium	0.5	5000	ND
	0.5	25000	ND
	0.5	50000	ND
	1.0	5000	ND
	1.0	25000	ND
	1.0	50000	ND
Lead	0.5	5000	ND
	0.5	25000	ND
	0.5	50000	0.1
	1.0	5000	0.2
	1.0	25000	0.1
	1.0	50000	3.5

ND = not detected
Source: [14]

The California Waste Extraction Test (WET) [4] differs from the traditional EPA EP Toxicity test in that the treated waste is ground to 200 mesh, then mixed with the extraction solution which consists of 0.2M sodium citrate. The mixed solution is adjusted to a pH of 5, and the extraction procedure lasts for forty eight hours instead of the twenty four hours in the EP procedure. The citric acid is a strong metal complexing agent and will generally leach more metals from a solidified waste than the acetic acid specified by the EP test.

Table 5.9. Multiple Extraction Data of a Cement-fixed Metal Sludge

	Metal Concentration in Leachate (mg/l)										Total Analysis of Sample	EP Toxicity
Metal	EP	1	2	3	4	5	6	7	8	9	mg/kg	Limits
Cd	.03	<.01	<.01	<.01	<.01	<.01	<.01	<.01	<.01	<.01	6.0	1.0
Hg	.0031	<.0002	<.0002	<.0002	<.0002	<.0011	<.0002	<.0002	<.0002	<.0012	1.0	0.2
Cr	.51	.22	.09	.19	.29	.35	<.02	<.02	<.02	<.02	1527.0	5.0
Pb	.04	.30	.22	.13	.16	.15	<.06	<.06	<.06	<.06	105.0	5.0
As	.006	<.001	<.001	<.001	<.001	<.003	<.001	<.001	<.002	<.001	.05	5.0
Se	.002	<.002	<.002	<.002	<.002	<.002	<.002	<.003	<.002	<.002	2.0	1.0

Source: [15]

5.4. SURVEY OF SOME EXISTING FIXATION PROCESSES

The principal difference among the chemical fixation processes offered by various companies is the different additives used to assist in fixing the wastes. Often, the same solidifying agent, for example cement or lime, may be used in two different processes, while the additives, usually proprietary in nature, are different. The result of such differences is that two processes which may appear similar might be applicable to different wastes. The proprietary additives might be incompatible with certain wastes, breaking down in their presence, and becoming ineffective for bonding. Thus, the type of waste that can be treated (fixed) is process-specific.

In this Section, we discuss some of the commercial cement- and lime-based fixation processes. These two techniques are by far the most widely used forms of fixation in the United

States. All of the information presented in this Section was obtained from either technical bulletins published by the companies or scientific papers. We present the data obtained from companies with regard to types of components treated and the results of strength and leaching tests. We believe that all such information must be viewed very critically and that prior to implementing a specific fixation technology, bench scale tests must be performed using the particular waste to be treated.

The CHEMFIX Process [16]

a. *Vendor's name and address:* Chemfix Technologies, Inc.
2424 Edenborn Ave.
Suite 620
Metairie, LA 70001
(504) 831-3600

b. *Category of fixing process*: Cement-based fixation system with soluble silicates acting as additives.

c. *Description of process and final product:* The reagents used in the CHEMFIX process react with polyvalent metal ions to produce stable and insoluble compounds. Other constituents are physically entrapped and immobilized within the matrix created during the reaction process. The end product is a clay-like material, having high unconfined compressive strength (2,000 to 10,000 lb/ft^2) and low permeability ($1x10^{-6}$ to $1x10^{-7}$ cm/sec). Typical volume increases are 5% to 10%.

d. *Types of wastes (chemically and physically) treated and their allowable limits:* The CHEMFIX process is able to handle heavy metal wastes, but the effectiveness is much reduced if concentrations of organic compounds are greater than 20% or if the concentration of oil or grease is greater than 10%. No documentation is available on effectiveness of treatment of high solvent concentration wastes. Table 5.10 lists the typical concentration limits of the components which can be treated, whether soluble or insoluble. The process can treat either high or low solid wastes. Wastewater or contaminated rain runoff with less than 1% solid content can be treated. If the pH of the waste is extremely low, a pretreatment such as lime is needed to neutralize the material. Likewise for an extremely alkaline waste, pretreatment to lower pH is necessary.

e. *Types of waste excluded from treatment:* The CHEMFIX process is not effective in the treatment of asbestos. The asbestos fibers cannot be bound within the product matrix.

Table 5.10. Waste Components and Their Concentration Limits for the CHEMFIX Process

Metals	Maximum Concentration (mg/kg)
Aluminum	100,000
Antimony	5,000
Arsenic	1,500
Barium	20,000
Beryllium	5,000
Cadmium	90,000
Chromium	80,000
Iron	150,000
Lead	100,000
Manganese	150,000
Mercury	1,000
Nickel	40,000
Selenium	2,000
Silver	10,000
Thallium	5,000
Zinc	250,000
Other Constituents	**Acceptable Ranges (mg/kg)**
Polynuclear Aromatic Hydrocarbons	<10,000
Oil and Grease	<300,000
Acids	3.5 - 11.5 (pH)
Semivolatile Organics	<10,000
Cyanide	<3,000

Source: [16]

f. *Cost of fixation*: Treatment costs range from $25 to $75 per ton, including bringing the equipment on-site, setting up the equipment, returning the equipment, operating personnel, and reagent costs.

g. *Leach and strength test:* The end product has an unconfined compressive strength between 2000 to 10,000 lb/ft^2. When used only for the wastes recommended for treatment, the final product can be certified as a non-hazardous material based on the EPA Extraction Procedure (EP) Toxicity Test. An example of the leaching test is shown in Table 5.11.

h. *Existing facilities*: Hyperion Sludge Treatment (Los Angeles, CA), Casmalia Resources (Casmalia, CA), Bailard Landfill (Oxnard, CA), South Essex Sewerage District (Massachusetts), Gloucester County Utilities Authority (New Jersey).

Table 5.11. Leaching Test of Incinerator Ash from Municipal Solid Wastes Using the CHEMFIX Process

Constituents	Conc. in Raw Waste mg/kg	CHEMFIX Product EP Extract (mg/l)	EP Toxicity Limits (mg/l)
Arsenic	14.95	< 0.002	5.0
Cadmium	8.50	0.13	1.0
Chromium	561.60	0.086	5.0
Lead	221.00	0.11	5.0
Mercury	90.00	0.002	0.2
Selenium	4.70	0.001	1.0
Silver	4.40	0.008	1.0

The Westinghouse Fixation Process [17]

a. *Vendor's name and address:* Westinghouse Electric Corp.
Environmental Technology Division
Box 286
Madison, PA 15663-0286
(301) 992-0066

b. *Fixation process:* Cement as solidifying agent with sodium silicate additives.

c. *Description of process and final product:* The sodium silicate additives increase the pH and precipitate metals as hydroxides. Portland cement is then used as the setting agent to bind the insoluble metal hydroxides within the cement matrix. The final product has a dry friable soil appearance, with compressive strength greater than 50 lb/in^2.

d. *Types of waste (chemically and physically) treated and their allowable limits:* The Westinghouse fixation process can treat wastes containing up to 50% organic compounds in the form of oil, slurries containing mixed soils, and wastes containing high concentrations of grit. For metal contaminated wastes, the process can handle arsenic, barium, cadmium, chromium, lead, mercury, selenium and silver.

e. *Types of waste excluded from treatment:* The process has not been used to treat waste containing asbestos and no conclusion can be made as far as the applicability for the treatment of asbestos wastes.

f. *Cost of Fixation:* Fixation costs are generally in the range of $30 to $80 per ton. The price would include mobilization, cost of disposable containers, usage of mixing equipment, cement, additive, labor and demobilization.

g. *Leach and strength tests*: When properly implemented, the final product has a compressive strength of more than 50 lb/in^2. It passes the EP Toxicity Test and can be regarded as non-hazardous.

h. *List of existing contracts:* Alabama Power Company, Arkansas Power & Light Co., Baltimore Gas & Electric Co., Commonwealth Edison Co., Consolidated Edison Co., Indiana & Michigan Electric Co., Iowa Electric Co., Long Island Lighting Co., Louisiana

Power & Light Co., Maine Yankee Atomic Power Co., Mississippi Power & Light Co., New York Power Authority, Northern States Power Co., Pennsylvania Power & Light Co., Public Service Electric & Gas, Rochester Gas & Electric Co., Tennessee Valley Authority, Virginia Electric Power Co., Wisconsin Electric Power Co., Yankee Atomic Power Co.

The CHEM-SECURE Process [18]

a. *Vendor's name and address:* Chemical Waste Management, Inc.
3001 Butterfield Rd.
Oak Brook, IL 60521
(312) 841-8360

b. *Category of fixing process:* Based primarily on inorganic, cementitious and pozzolanic reagents.

c. *Description of process and final product:* This process is often used for off-site treatment of wastes. The waste is first transported to the treatment facility. It is then combined with the solidifying reagent in an enclosed mechanical mixer and allowed to set, entrapping the contaminants within the solid matrix. After stabilization, the end product is a solid with a bearing strength of at least 2,000 lb/ft^2 minimum. The volume increase varies from less than 10% to about 50%, depending upon the waste composition. Permeability also varies, but is usually less than 10^{-5} cm/sec.

d. *Types of waste (chemically and physically) treated and their allowable limits:* Waste with any water or solid content can theoretically be treated. The vendor claims that the process is able to solidify wastes with any organic content.

e. *Types of waste excluded from treatment:* Wastes containing asbestos.

f. *Cost of fixation:* Costs vary widely with waste type, quantity, location and final disposal decision. The vendor will quote fixed price or unit cost for the complete stabilization operation, whether performed on-site or at a Chemical Waste Management facility.

g. *Leach and strength tests:* The product has a bearing strength of at least 2000 lb/ft^2. The leaching test results meet the required limits as set forth by the EPA Toxicity Test.

h. *List of existing contracts of facilities:* Kettleman Hills, CA; Lake Charles, LA; Emelle, Al; Model City, NY; Calumet City, IL.

The SRS/EIF Oily Sludge Fixation Process [19]

a. *Vendor's name and address:* Mittelhauser Corporation
23272 Mill Creek Road, Suite 300
Laguna Hills, CA 92653
(714) 472-2444

b. *Category of fixation process:* Lime-based process with proprietary non-toxic chemicals that catalyze and control the reactions between the lime and the waste.

c. *Description of process and final product:* The sludge to be treated is placed in a blending pit. Specially prepared lime with additives is added to the sludge in the blending pit using an excavator. The first step of the process, neutralization, changes the sludge to a grey paste and disperses any solid pieces present. After fifteen minutes, a second lime preparation is added, which has a chemical composition that is different from the first batch of lime. The lime is mixed over a twenty minute period in the blending pit. After this time, the reaction is almost complete. The stabilized waste is then allowed to cure for four weeks. The final product is a dirt or clay-like material with good compactive properties. The permeability of the compacted material can be less than 10^{-12} cm/sec. The volume change upon treatment is typically 25 to 35% and compressive strengths as high as 88 lb/in^2 have been measured.

d. *Types of waste (chemically or physically) treated and their allowable limits:* The SRS/EIF process can treat oily sludges that contain a minimum of 3% up to 80% organic compounds. The types of organics that have been treated include: crude oil, refinery intermediates or final products, polychlorinated biphenyls, pesticides, sludges, tars, painting wastes, and acid sludges. It can also be used to treat wastes containing lead, antimony, arsenic, barium, boron, cadmium chromium, copper, mercury, nickel, selenium, silver and zinc at concentrations up to 3000 ppm. There is no limit on the water or solid content of the waste and all ranges of pH can be treated.

e. *Types of waste excluded from treatment:* The process has not been used to treat wastes containing asbestos.

f. *Cost of fixation:* The average price for processing a cubic yard of waste is $65 to $90, depending on the characteristics of the material to be stabilized. The labor, maintenance and other related costs are included in this price.

g. *Leach and strength tests:* Treated wastes have been tested using both the Liquid Release Test and the Unconfined Compressive Strength (ASTM D-2166) Test. The end product has a compressive strength as high as 88 lb/in^2. The leaching rates of contaminants as determined by the Liquid Release Test are all within the toxicity limits specified by the Environmental Protection Agency.

h. *List of existing contracts or facilities:* Sand Springs Petrochemical Complex, Sand Springs, OK, and Oklahoma State Dept. of Health, Oklahoma City, OK.

The ENSOL Process [20]

a. *Vendor's name and address:* Ensotech, Inc.
11300 Hartland St.
North Hollywood, CA 91605
(818) 760-8622

b. *Category of fixation process:* Lime-based process that uses an anionic polymer and a chelating agent as additives.

c. *Description of process and final product:* In this two step process, the anionic polymer first acts to attract metal ions. In the second step, the chelating agent reacts with the metal, bonding with it to render it unleachable. Lime is then added to the mixture as a setting agent, forming the final product: a monolithic solid.

d. *Types of waste (chemically or physically) treated and their allowable limits:* The ENSOL process is useful in fixing all heavy metals, such as, lead, copper, zinc, cadmium, cobalt, selenium, mercury, silver, chromium and nickel. In addition, it reduces Cr^{+6} to Cr^{+3}, and reduces Se^{+6} to Se^{+4}. Heavy metal concentrations several times higher than the EPA Leach Test limits can be reduced to safety limits. The process is applicable to solids and sludges containing up to 70% water over a wide pH range (4 to 12).

e. *Types of waste excluded from treatment:* Pretreatment of waste is required if organic compounds and cyanides are present. Strong oxidants, for example, concentrated acids particularly nitric acid, which can react violently with the additives, are not suitable. The process has not been used to treat wastes containing asbestos.

f. *Cost of fixation:* Based upon approximately 200 cubic yards of material to be treated, the cost is $50 to $80 per cubic yard if the total metal concentrations are less than 2000 ppm, and $90 to $110 per cubic yard if the metal concentrations are up to 10,000 ppm. The costs consist of: capital - 20%, labor - 30%, maintenance - 5% and raw materials (lime, chemical additives) - 45%.

g. *Leach and strength tests:* The final product passes the EPA Leach Test and California Waste Extraction Test. When properly used, the final solid product can be classified as a nonhazardous waste and disposed of in a Class III (sanitary) landfill.

h. *List of existing facilities:* Ensotech, Inc., Sun Valley, CA 91352.

5.5. ASSESSING THE TREATABILITY OF CALIFORNIA WASTES BY CHEMICAL FIXATION

As seen from the discussion in section 5.4, the characteristics of the waste that can be treated by the various processes can vary tremendously. For example, the CHEMFIX process can only treat wastes containing up to 1% organic compounds, whereas the CHEM-SECURE process claims to be capable of handling wastes with any organic content. With the exception of the CHEMFIX process, the vendors do not generally report the treatability limits on metal fixation in detail, nor do they publish data on leaching tests. This lack of information severely limits our ability to determine the applicability of any of these treatment technologies to a wide spectrum of wastes. However, after detailed discussions with engineers at Chemfix Technology Co., and also with personnel associated with the regulatory branch of government who have detailed knowledge of chemical fixation technology, we concluded that the treatability limits for the CHEMFIX process given in Table 5.10 are quite characteristic of a fixation process. We used these limits in this study to explore the applicability of chemical fixation technologies to our model spectrum of wastes based upon the 1985 Biennial Generator Report database [21]. The treatability limits in Table 5.10 have been converted to more appropriate units, percent of waste, as shown in Table 5.12. The flow chart for the treatability determination and cost estimation is given in Figure 5.1, and the results are shown in Table 5.13. A description of the program and the information generated from it is discussed in the remainder of this Section.

Table 5.12. Waste Components and Their Treatability Limits Used in Chemical Fixation Model Calculations.

Component/ Chem. Formula	Treatability Limit (%)
Aluminum	10.0
Antimony	0.5
Arsenic	0.15
Barium	2.0
Beryllium	0.5
Cadmium	9.0
Chromium	8.0
Iron	15.0
Lead	10.0
Manganese	15.0
Mercury	0.1
Nickel	4.0
Selenium	0.2
Silver	1.0
Thallium	0.5
Zinc	25.0
Organics	1.0
Cyanide	0.3

Source: [16]

Table 5.13. Results on Treatability Using Chemical Fixation

CAL	SIC	VOLUME (TONS)	FIXATION COST ($)	GENERATING PROCESS
0	99	0.50	25.0	Lab experimentation
111	34	2.71	135.5	Conversion coating aluminum/sodium hydroxide
111	36	12.53	626.5	Semiconductor process
111	36	5.96	298.0	Printed circuit mfg. process
111	36	1.38	69.0	Printed circuit mfg. process
111	36	.92	46.0	Printed circuit mfg process
111	34	4.00	200.0	Chemical milling
111	34	46.00	2,300.0	Chemical milling
111	37	3.00	150.0	Manufacture and testing of rocket engine part
111	37	1.00	50.0	Nickel base alloy cleaning
111	37	0.90	45.0	Metals cleaning
111	34	41.67	2,083.5	Metal surface processing
111	34	12.50	625.0	Aluminum deoxidation
111	36	1.04	52.0	Semiconductor wafer fabrication
111	34	403.30	20,165.0	Photo chem. machining (etching) of thin metal sheets
111	34	5.00	250.0	Waste aqueous sulfuric acid anodizing solution
111	34	1.67	83.5	Metal treatment
111	34	4.17	208.5	Spent acid activating solutions from electroplat. operations
111	34	87.50	4,375.0	Anodizing of aluminum
111	39	225.00	11,250.0	Cleaning of brass and copper
111	34	1.88	94.0	Spent acid from strip tank
111	34	12.50	625.0	Anodizing solution and acid pickle
111	34	19.55	977.5	Site clean-up materials
111	36	193.00	9650.0	Waste acid baths from chemical milling process
111	34	0.83	41.5	Alum. conversion coating/washer has retain wall to hold spills
111	36	3.00	150.0	PC board mfg.
111	34	275.00	13,750.0	Steel pickling
111	36	4.58	229.0	Metal plating baths
111	34	25,000.00	1,250,000.0	Electroplating, copper, nickel, chrome, cyanide
111	34	60.42	3,021.0	Treatment baths
111	34	5.00	250.0	Plating shop
111	34	16.26	813.0	Nitric acid strip
111	36	63.50	3,175.0	Printed circuit board mfg.
111	39	0.44	22.0	Acid baths for parts cleaning
111	39	0.50	25.0	Acid baths
111	37	27.00	1,350.0	Metal cleaning
111	34	18.00	900.0	Deoxidizing of aluminum prior to painting
111	32	10.50	525.0	Drag out from electroplating tank, cleanup of ceramic clay/material
111	36	2.52	126.0	Copper conditioner
111	36	30.71	1,535.5	Solder plating solution and metal surface cleaner
111	36	0.70	35.0	Sludge from plating solution
111	99	3.15	157.5	Metal finishing
111	99	0.20	10.0	Circuit board shop
111	99	0.20	10.0	Metal finishing
111	97	1.83	91.5	Plating shop operations

(continued)

Table 5.13. (continued)

111	97	47.64	2,382.0	Plating shop operations
111	39	0.12	6.0	Circuit card etching
111	39	0.12	6.0	Metal plating
111	39	0.12	6.0	Circuit card etchant
111	39	0.12	6.0	Metal plating
111	39	0.06	3.0	Copper etching
111	39	0.25	12.5	Circuit board etching
111	39	0.06	3.0	Circuit card treatment
111	36	700.00	35,000.0	Mfg of printed circuit boards
111	99	30.00	1,500.0	Plating rinse
111	39	102.00	5,100.0	Drag out tanks from electroplating
111	34	1.60	80.0	Etching and plating of printed circuit boards
111	34	9.32	466.0	Aluminum anodizing
111	36	110.00	5500.0	Plating
111	36	3.90	195.0	Wafer preparation
111	36	6.50	325.0	Semiconductor mfg.
111	30	35.00	1750.0	Chrome electroplating operation
111	39	12.50	625.0	Chrome/nickel electroplating
111	37	6.00	300.0	Metal finishing
111	34	230.42	11521.0	Waste plating baths or waste rinse baths
111	39	0.23	11.5	Circuit board etching solution
111	36	2.52	126.0	Servicing and adjustment lead acid batteries
111	28	11.23	561.5	Plating rack stripper
111	28	8.44	422.0	Copper etching
111	28	10.00	500.0	Spent mixed chemicals
111	37	15.00	750.0	Misc. acid waste from labs
111	36	6.80	340.0	Industrial battery repair and replacement
111	73	8.50	425.0	Anodizing
111	39	26.00	1300.0	Etching copper using ferric chloride
111	36	16.00	800.0	Printed circuit board mfg.
111	36	899.17	44958.5	Spent hydrofluoric acid solution from semiconduct. mfg
111	39	0.90	45.0	Textile coating
111	99	0.46	23.0	Printed wiring board production
111	37	1.25	62.5	Plating
111	37	42.50	2125.0	Plating
111	37	10.00	500.0	Plating - rack stripping
111	39	14.58	729.0	Spent acid solution used in plating process
111	28	54.36	2718.0	Acid etch rinse water and spent baths
111	36	6.15	307.5	Electronic processing
111	36	8.33	416.5	Drag out from barrel plating and copper etching process
111	49	0.42	21.0	Cleaning of boilers
111	49	73.10	3655.0	Cleaning of boiler
111	95	32.71	1635.5	Spent solutions from electroplating
111	99	6.25	312.5	Plating waste
111	99	10.42	521.0	Tank clean up
111	39	9.17	458.5	Plating waste
111	36	6.42	321.0	Glass cleaning
111	37	322.00	16100.0	Deoxidizing aluminum alloys
111	37	15.00	750.0	Prebond etching of aluminum
111	37	0.12	6.0	Spoiled chromic acid mixture lab packed
111	37	131.08	6554.0	Metal finishing operations
111	37	5.00	250.0	Cleaning of surplused process tanks
111	37	103.00	5150.0	Deoxidizing aluminum alloys

(continued)

Table 5.13. (continued)

111	37	11.00	550.0	Process bath that applies chem film coating to aluminum
111	37	48.00	2400.0	Deoxidizer treatment of aluminum
111	34	18.00	900.0	Cleaning and coating process for aluminum
111	37	63.00	3150.0	Metal cleaning/stripping
111	37	15.00	750.0	Metal cleaning
111	37	52.50	2625.0	Plating
111	28	229.17	11458.5	Metal pickling
111	36	37.00	1850.0	Plating processes
111	27	3.44	172.0	Electrostatic precipitation
111	97	47.64	2382.0	Plating shop mfg.
111	37	17.00	850.0	Acidic waste water treatment liquid prior to filter pressing
111	37	29.00	1450.0	Acids from circuit board mfg.
111	28	41.81	2090.5	Spent pickling liquors & spent plating baths
111	28	1.50	75.0	Aircraft parts mfg.
111	48	4.70	235.0	Central office removal
111	48	0.40	20.0	Central office removal
111	28	8.33	416.5	Chrome plating
111	99	97.92	4896.0	Aluminum anodizing
111	28	30.00	1500.0	Mfg. of photographic & electronic chemicals
111	36	11.46	573.0	Printed circuit board mfg.
111	36	3.40	170.0	Metal stripping
111	36	0.50	25.0	Metal stripping
111	37	0.04	2.0	Plating of small electronic components
111	35	8.33	416.5	Chromic acid plating
111	37	5.00	250.0	Spent color chem film for aluminum mixed w/spent HCl solution
111	37	4.50	225.0	Spent dichromate sealing sol w/azo dye contamination
111	37	19.00	950.0	Spent pre-bond aluminum etch solution
111	37	5.00	250.0	Spent aluminum deoxidizing solution
111	37	125.00	6250.0	Tubing fabrication
111	37	6.25	312.5	Conversion coating for aluminum
111	37	4.17	208.5	Cr inhibited pickle for titanium alloy
111	36	3.12	156.0	Spent circuit board etching solution
111	36	1.20	60.0	Cleaning chrome tank
111	95	10.42	521.0	Printed circuit board acid waste
111	34	1.67	83.5	Stripping and etching operations
111	36	27.04	1352.0	Printed circuit mfg. process
111	36	0.69	34.5	Printed circuit mfg. process
111	36	2.00	100.0	PC board mfg.
111	73	0.46	23.0	Pesticide R&D
111	36	38.80	1940.0	Electroplating circuit boards
111	34	15.00	750.0	Electroplating
111	36	41.67	2083.5	Printed circuit mfg.
111	37	1.04	52.0	Chemical treatment of aluminum parts
111	37	0.23	11.5	Unknown 55 gal. drum contents
111	36	1.48	74.0	Stripping photoresist from wafer & mask cleaning
111	36	7.50	375.0	Metal cleaning & electroplating
111	36	2.30	115.0	Copper, lead, tin stripping
111	36	1.88	94.0	Copper electroplating operation
111	36	4.00	200.0	Ferric chloride
111	99	13.75	687.5	Spent chrome-sulfuric solution
111	99	8.75	437.5	Spent plating pretreatment acids and drag outs

(continued)

Table 5.13. (continued)

111	82	3.30	165.0	Lab glassware cleaning
111	28	1.09	54.5	Anodizing prep. solutions
111	36	83.33	4166.5	Electroplating/cleaning of electronic parts
111	34	3.00	150.0	Spill
111	34	421.67	21083.5	Zinc electroplating and zinc phosphatizing
111	38	2.29	114.5	Aluminum finishing/cleanup with absorbent
111	99	312.50	15625.0	Steel pickling bath-neutralized HCl with pH>5
111	39	4.00	200.0	Copper etching
111	39	2.00	100.0	Plating waste
111	36	125.00	6250.0	Metal finishing
112	28	6.42	321.0	Compounding of reagent
112	36	6.64	332.0	Computer disc coating
112	29	9.58	479.0	Chemical cleaning
112	36	510.40	25520.0	Wafer preparation
112	39	5.31	265.5	Spill cleanup
112	37	168.00	8400.0	Plating solution
112	30	3.75	187.5	Methoxylation of chlorosilanes
112	28	716.00	35800.0	Pesticide (Devrinol) production
112	36	0.35	17.5	Electrical and electronic machinery
113	36	3.00	150.0	Removal of lead/tin from circuit boards
113	28	1.35	67.5	Storage residues
113	34	1500.00	75000.0	Steel pickling and zinc removal
113	97	13.58	679.0	Plating shop operations
113	36	282.00	14100.0	Waste acid from I.C. wafer etching
113	36	1.30	65.0	Semiconductor assembly-testing
113	36	7.67	383.5	Semiconductor testing
113	34	2.00	100.0	Chemical etching of metals
113	37	76.25	3812.5	Metal finishing operations
113	37	32.00	1600.0	Cleaning of surplused process tank
113	28	130.00	6500.0	Commerc. byprod. (from hydrolysis of spent alum. chloride cat.)
113	36	22.00	1100.0	Spent acidic sol. from cleaning of aluminum
113	37	10.00	500.0	Spent titanium etch & steel passivation soln.
113	37	1.50	75.0	Spent aluminum and oxidize strip soln.
113	37	166.32	8316.0	Aluminum cleaning
113	36	2.29	114.5	Circuit board mfg.
113	28	600.00	30000.0	Etching, lab chemicals
113	28	0.13	6.5	Lab chemicals off spec.
121	36	1.83	91.5	Printed circuit mfg. process
121	34	2845.00	142250.0	Chemical milling
121	34	41.86	2093.0	Sulfuric acid aluminum anodizing
121	34	146.00	7300.0	Chemical etching of aluminum
121	39	0.04	2.0	Copper plating
121	34	41.74	2087.0	From industrial waste incinerator
121	28	250.00	12500.0	Printing ink mfg.
121	37	175.00	8750.0	Chemical milling of aluminum
121	34	36.00	1800.0	Treatment of aluminum prior to painting
121	34	170.00	8500.0	Spent electroplating, stripping & cleaning bath solns.
121	07	260.87	13043.5	Plating shop operations
121	97	42.52	2126.0	Plating shop operations
121	37	1458.33	72916.5	Aluminum milling of aircraft parts
121	28	37.00	1850.0	Plating, nickel chrome
121	37	65.00	3250.0	Cleaning soln. for locomotive turbos & traction motors
121	39	25.03	1251.5	Stripping operation

(continued)

Table 5.13. (continued)

121	37	50.00	2500.0	Paint stripping
121	38	421.00	21050.0	Metal parts cleaning and anodizing waste
121	37	6.25	312.5	Plating
121	32	17.00	850.0	Glass stripping ceramic, screen cleaning
121	97	42.52	2126.0	Plating shop operation
121	34	12.50	625.0	Sludge from bottom of clarifier of anodize & etch dept.
121	36	0.92	46.0	Printed circuit mfg. process
121	24	25.00	1250.0	Site cleanup
121	36	237.00	11850.0	Electroplating
121	36	2.57	128.5	Spent plating bath solution
121	40	82.08	4104.0	Railcar parts cleaning
121	39	5.00	250.0	Electroless copper plating of PC boards
121	39	0.52	26.0	Stripping process sodium hydroxide & $NaNO_3$
121	13	7.27	363.5	Caustic/water flood project shut-down
121	99	41.67	2083.5	Steel pickling bath - ammonium chloride solution
121	99	50.00	2500.0	Steel pickling bath - alkaline solution with metals
121	99	83.33	4166.5	Steel pickling bath - sodium hydroxide solution
122	29	600.00	30000.0	Naptha and kerosene caustic treating
122	29	109.76	5488.0	Chemical cleaning
122	29	4.58	229.0	Chemical cleaning
122	29	5.42	271.0	Chemical cleaning
122	28	25065.00	1253250.0	Process scrubbers
122	99	22.92	1146.0	Tank cleaning
122	29	1000.00	50000.0	Spent caustic from alkylation operations
122	75	42.00	2100.0	Cleaning of auto parts
122	34	17.00	850.0	Electroplating operations
122	28	26.60	1330.0	Industrial sewer sludge
122	95	3.40	170.0	Water and sewage treatment
122	28	131.20	6560.0	Pharmaceutical intermediate mfg.
122	36	1.00	50.0	Sodium hydroxide used in solvent recovery
122	36	1.00	50.0	Sodium hydroxide used in solvent recovery
122	99	13.90	695.0	Cleaning of metals for the penetrant inspection process
122	39	10.42	521.0	Ink stripping
123	34	115.00	5750.0	Cleaners and sludge from electroplating
123	28	16.67	833.5	Caustic solution for cleaning portable paint vats
123	34	200.00	10000.0	Drum reconditioning
123	34	550.00	27500.0	Drum reconditioning
123	28	12007.00	600350.0	Process wash down water and lab drains
123	32	289.40	14470.0	Glass fabrication, mirror mfg.
123	99	2.92	146.0	Plating waste
123	37	13.50	675.0	Rinse water tanks from plastic shop hotmelt process
123	50	294.00	14700.0	Drum reconditioning
123	29	200.40	10020.0	Claus plant quench tower
123	28	59.00	2950.0	Manufacture of photographic & electronic chemicals
123	39	20.83	1041.5	Electroplating stripping & cleaning solution
123	36	9.00	450.0	Spent alkaline cleaning solutions from aluminum processing
123	34	53.00	2650.0	Aluminum anodizing processes
123	33	24.00	1200.0	NaOH solution used to clean tooling
123	37	12.50	625.0	Metal parts cleaning
123	37	1310.00	65500.0	Plating shop waste
131	37	505.40	25270.0	Removal of propellant from rocket cases w/high pressure H_2O

(continued)

Table 5.13. (continued)

131	28	4.97	248.5	Corrosion inhibition of well
131	36	0.48	24.0	Electrical and electronic machinery
131	13	1158.00	57900.0	Oilwell stimulation
131	36	2.75	137.5	Silver plating process for copper bus bar
131	36	12.00	600.0	Clean-up of clarifier
131	37	20.00	1000.0	Metal finishing activities
131	36	3.00	150.0	Silicon wafer etching and quartz etching
131	28	57.08	2854.0	Mixed chemicals from plating process & rinse water
131	36	40.00	2000.0	Aluminum chemical conversion coating
131	34	5.00	250.0	Electroplating operations
131	28	3.50	175.0	Cleaning metal parts
131	97	0.10	5.0	Laboratory cleanup
131	36	306.67	15333.5	Spent plating and cleaning baths
131	39	1.55	77.5	Spent plating and stripping solution
131	39	0.28	14.0	Empty drums: last contained corrosive materials
131	36	1.15	57.5	Semiconductor mfg.
131	10	25.00	1250.0	Metallurgical pilot plant tailings
132	34	49.50	2475.0	Metal plating
132	34	157.00	7850.0	Chemical milling
132	34	104.00	5200.0	Salts from the chemical milling metals
132	34	101.25	5062.5	Metal surface processing
132	28	129.17	6458.5	Sludge from waste water treatment
132	34	23.80	1190.0	Waste water pretreatment
132	34	15.00	750.0	Electroless nickel plating
132	32	25.00	1250.0	Clarifier sludge
132	28	32.92	1646.0	Drum residual/blending tank rinsing
132	13	4538.00	226900.0	Stretford process
132	34	12.50	625.0	Process sewer line
132	36	0.23	11.5	Metal plating tank
132	35	18.75	937.5	Metal preparation
132	99	22.29	1114.5	Chemical neutralization and waste water
132	34	370.00	18500.0	Wastewater chlorination
132	28	47.08	2354.0	Clean-up material from plastics compounding
132	34	6.25	312.5	Spent acid pickles
132	34	50.00	2500.0	Treatment baths
132	39	16.67	833.5	Aluminum etching tank
132	39	0.42	21.0	Clean and rinse stages of pre-plant washer for steel
132	39	20.83	1041.5	Waste paint pre-treatment solution for steel
132	29	2.29	114.5	Petroleum research
132	29	1000.00	50000.0	Effluent treating plant
132	36	13.29	664.5	Printed circuit mfg.
132	34	30.00	1500.0	Absorbent material
132	34	410.00	20500.0	Treatment of aluminum prior to painting
132	36	10.76	538.0	Site cleanup material
132	97	1.31	65.5	Plating shop operations
132	97	18.75	937.5	Plating shop operation (chromate conversion)
132	39	227.69	11384.5	Spent plating bath solutions
132	99	12.50	625.0	Manufacture of printed circuit boards
132	36	3.21	160.5	Decommissioning R&D electroplating lab
132	99	337.50	16875.0	Anodizing and chemical milling
132	29	2000.00	100000.0	Stretford froth from tail gas clean up operations
132	36	0.69	34.5	Plating
132	37	118.00	5900.0	Auto body washing operations
132	37	87.00	4350.0	Semi-annual clarifier cleaning

(continued)

Table 5.13. (continued)

132	34	10.42	521.0	Plating, cleaning
132	34	603.93	30196.5	Back washing of resin cylinders
132	37	4.17	208.5	Clarifier
132	37	2.50	125.0	Aluminum dip braze
132	36	200.00	10000.0	Sludge generated from spent plating baths
132	36	125.00	6250.0	Manufacture of memory discs for computers
132	36	0.60	30.0	Phosphate process tank for phosphating cadmium-plating
132	33	304.00	15200.0	Solution from etching process
132	34	27.92	1396.0	Manufacture hardware for computer peripherals
132	32	18.00	900.0	Nickel plating
132	37	379.33	18966.5	Metal finishing operations
132	34	44.00	2200.0	Coating process for aluminum
132	36	46.00	2300.0	Electroplating electronic components
132	28	10.00	500.0	Cleanup chromium prod. blender reduction w/sodium bisulfate
132	34	2.10	105.0	Copper strip processing
132	34	0.69	34.5	Metal parts cleaning
132	34	17.50	875.0	Nickel sulfamate electroforming, rhodium electroplating, chrome
132	36	84.58	4229.0	Clear anodizing waste spill pillows
132	99	41.67	2083.5	Recycling jewelry waste
132	37	16.67	833.5	Paint sludge from water wash spray booth
132	37	166.67	8333.5	Cleaning of tank overflow containment
132	37	91666.67	4583334.0	Surface treatment of metals
132	36	75.00	3750.0	Electroplating waste water treatment w/sludge
132	95	10.42	521.0	Plating shop
132	36	311.29	15564.5	Printed circuit mfg. waste water treatment
132	13	215.00	10750.0	SO_2 scrubber stack wash water
132	28	120.70	6035.0	Water from washing equipment in catalyst mfg.
132	28	489.50	24475.0	Floor/equipment washing
132	39	5833.33	291666.5	Neutralization of waste acids/caustics generated by industry
132	34	2.00	100.0	Plating and anodizing of metals - clarifier waste
132	36	11.80	590.0	Manufacture of PC boards
132	33	150.00	7500.0	Treatment of magnesium dust
132	34	2.94	147.0	Waste water from plating shop sump cleanup
132	13	85.10	4255.0	Oil production waste
132	39	8.33	416.5	Nickel plating operation
132	34	2.83	141.5	Chemical coating wash down
132	34	8.30	415.0	Etching and plating of PC boards
132	28	16.00	800.0	Vessel washings from mfg. of acid cleaning products
132	29	4519.00	225950.0	Stretford solution - refinery sulfur plants purge
132	34	229.17	11458.5	Oil, flux, floatable material skimmed from neutralizer
132	36	1000.00	50000.0	Spent liquid mixture from electroplating operations
132	36	3563.00	178150.0	Metal finishing
134	28	1040.00	52000.0	Manufacture of 3-nitro-1,5-pentane diisocyanate
134	28	15145.00	757250.0	Manufacture of nitroplasticizers

(continued)

Table 5.13. (continued)

134	27	110.42	5521.0	Cleaning press parts and ink pumps
134	28	106.86	5343.0	Water washes and sludges from cleaning tubs & equipment
134	35	54.38	2719.0	Monitoring well sampling and testing
134	35	60.42	3021.0	Monitoring wells pumping, testing
134	99	0.46	23.0	Mixed solvent/cleanup material
134	36	16.43	821.5	Chemical storage tank cleaning
134	34	17.92	896.0	Metal cleaning
134	28	8.33	416.5	Cleanup of water-based ink production area & equipment
134	48	11.67	583.5	Waste water removed from a vehicle steam cleaning sump
134	48	389.00	19450.0	Leaking underground storage tank
134	36	293.00	14650.0	Aluminum conversion/paint shop
134	28	25.00	1250.0	Solution coating of synthetic substrates
134	28	15.00	750.0	Synthetic resin impregnation of synthetic substrates
134	39	12.00	600.0	Deburring of aluminum castings
134	75	10.00	500.0	Steam cleaning residue - mud & water
134	37	30.53	1526.5	Manufacturing operations
134	28	150.00	7500.0	Polymer mfg.
134	37	280.00	14000.0	Spill cleanup
134	37	250.00	12500.0	Clarifiers serving water wash paint booths, steam cleaner
134	37	241.67	12083.5	Waste coolant from machining of parts
134	99	10.80	540.0	Rainfall into contaminated excavation
134	20	67.00	3350.0	Wastewater basin sediment
134	37	48.80	2440.0	Cleaning of waste tank
134	28	7.50	375.0	Water washes and sludges from cleaning tubs & equipment
134	28	12.50	625.0	Adhesive mixer wash
135	99	1.04	52.0	Rainwater collected in drums previously containing trichloroeth.
135	28	105.00	5250.0	Tank bottom from settlings tank of the total plant effluent
135	38	0.40	20.0	Etching process
135	30	33.33	1666.5	Factory floor washdown
135	29	600.00	30000.0	Spent Stretford solution
135	29	30.00	1500.0	Stretford (sulfur plant) froth
135	37	55.00	2750.0	Etching steel
135	28	6.00	300.0	Demolition of refrigeration equipment
135	28	398.00	19900.0	Pharmaceutical process wash water
135	28	16725.00	836250.0	Waste water from agricul. chem. production
135	36	0.06	3.0	Circuit board mfg.
135	99	528.00	26400.0	Fire and water damage/site cleanup
135	48	6.25	312.5	Tank wash out before removal
135	36	0.50	25.0	Ground well sludge
135	37	0.69	34.5	Sand blasting
135	36	1.22	61.0	Cadmium plating bath
135	39	1.25	62.5	Spill cleanup
135	36	1.04	52.0	Grinding/polishing
35	34	6.00	300.0	Burnishing operations
135	37	18.00	900.0	Ultrasonic verification testing
135	29	6330.00	316500.0	Stretford solution from SRU

(continued)

Table 5.13. (continued)

135	29	2469.00	123450.0	Removing Stretford solution from SRU unit & FCC CAT fines
135	48	5.00	250.0	Sulfuric acid spill cleanup
135	49	263.70	13185.0	Draining cooling water systems
135	13	131.00	6550.0	Gasoline/gas separating facility
135	34	76.00	3800.0	Non-cyanide grinding sludge from 1018 steel ball grinding tank
135	37	12.50	625.0	Anodize-metal parts
135	37	196.67	9833.5	Aluminum deoxidizer
135	13	142.00	7100.0	Oil production operations
135	13	6.00	300.0	Cleaning of water filtration equipment
135	34	2.41	120.5	Metal cleaning and prep/soap and water
135	29	277.00	13850.0	Contaminated water cleanup
135	28	2.00	100.0	Mfg. & packaging of powdered alkaline cleaners
135	82	10.00	500.0	Neutralized etching solution from experiment. production facil.
135	29	44.00	2200.0	Cooling tower basin cleaning
141	99	0.17	8.5	Metal finishing
141	99	0.20	10.0	Paint shop
141	36	0.50	25.0	Circuit board mfg.
141	38	0.50	25.0	Plating shop empty cans
141	39	0.11	5.5	Chemical lab/various production process
141	39	11.13	556.5	Mfg. processes & chemical lab
141	36	0.25	12.5	Obsolete lab chemicals
141	28	2.00	100.0	Mfg. and sale of specialty chemicals
141	37	2.40	120.0	Aerospace research and development
141	29	10.00	500.0	Disposal of contaminated process chemical
141	36	0.30	15.0	Micro-electronic lab chemicals
141	36	0.23	11.5	Chemicals purchased for electronic research
161	29	2500.00	125000.0	Refinery processing
161	29	1300.00	65000.0	Petroleum refining - precipitator fines
162	29	5.00	250.0	Refinery process
162	28	452.00	22600.0	Site cleanup material
162	29	225.80	11290.0	Spent catalyst disposal
162	28	2.00	100.0	Sulfuric acid plant
171	34	72.00	3600.0	Chemical milling
171	34	199.10	9955.0	Wastewater clarifier sludges & neutralized pickling acids
171	34	3.20	160.0	Sludge from on-site treatment of wastewater from electroplating
171	99	5.00	250.0	Zinc metal plating
171	34	0.75	37.5	Non-cyanide plating bath metal sludge
171	36	417.07	20853.5	Waste treatment from PC mfg.
171	36	57.52	2876.0	Site cleanup material
171	39	7.88	394.0	Zinc sludge from polishing die-cast parts
171	28	184.00	9200.0	Plating, nickel chrome
171	99	35.00	1750.0	Site cleanup material
171	34	74.62	3731.0	Porcelain mfg.
171	36	5.83	291.5	Sludge from the wastewater treatment system
171	36	1560.00	78000.0	Pretreatment
171	34	90.00	4500.0	Electroplating
171	37	0.50	25.0	Filter cake, chrome destruct
171	42	0.08	4.0	Spill

(continued)

Table 5.13. (continued)

171	34	60.00	3000.0	Pretreatment for sewering of metal finishing effluent
171	39	4.20	210.0	Clarifier sludge
171	34	382.00	19100.0	Electroplating operations
171	37	2.00	100.0	Waste rinse water treated to remove cromium as sludge
171	35	135.00	6750.0	Manufacturing of stoves
171	28	1.00	50.0	Repairs/equipment dismantling
171	36	21.00	1050.0	Sludge generated from waste water pretreatment facility
171	36	8.00	400.0	Mfg. of printed circuits, waste treatment cake
171	28	73.00	3650.0	Filtration/dewatering of mfg. wash wastes
171	49	17.20	860.0	Wastewater filtering waste
171	36	1.00	50.0	Polishing crystals
171	34	299.00	14950.0	Non-cyanide alkaline zinc plating on carbon steel
171	34	15.00	750.0	Electroplating
171	39	1.46	73.0	Nitric acid stripping of copper, tin, & lead
171	34	360.00	18000.0	Waste water treatment sludge from zinc/steel electroplating
171	36	20.00	1000.0	Metal finishing
172	37	80.00	4000.0	Machining of rocket engine parts
172	42	3.00	150.0	Boiler fuel-crude oil combustion
172	36	5.00	250.0	Old plating tanks, fans, ducts, scrubbers, treatment tanks
181	39	40.00	2000.0	Dewater sludge (filter press) from pretreatment of rinse water
181	28	8.60	430.0	Spent electroplating chemistry
181	34	9.50	475.0	Solder pot skimmings
181	34	26.62	1331.0	Metal hydroxide sludge
181	28	125.00	6250.0	Electronics mfg.
181	29	55.00	2750.0	Contaminated dehydrant
181	29	70.00	3500.0	Petroleum refining
181	34	4.00	200.0	Manufacture of steel cans
181	28	0.56	28.0	Cleaning of zinc additive system
181	36	1.65	82.5	Wafer preparation
181	34	2.25	112.5	Metal cleaning
181	28	1.50	75.0	Wet floor neutralization
181	39	12.00	600.0	Cleaning line
181	38	2.00	100.0	Chemical spill training
181	36	2.00	100.0	Scrubber waste
181	36	0.01	0.5	Integrated circuit mfg.
181	34	1.00	50.0	Electroplating plating cleaning process
181	36	3.00	150.0	Sludge from treatment of wastewaters/sweeper
181	39	0.90	45.0	Plating tank cleanups
181	39	0.23	11.5	Metal cleaning tank cleanup
181	37	50000.00	2500000.0	Shredded car bodies
181	99	0.03	1.5	Site cleanup material
181	28	7.08	354.0	Spill cleanup material
181	49	0.85	42.5	Degreasing electrical equipment parts
181	37	2.75	137.5	Sludge from conversion coating process tanks
181	37	1.25	62.5	Paint stripping, conversion coating, cleaning process
181	37	1.25	62.5	Clarifier sludge

(continued)

Table 5.13. (continued)

181	37	1.12	56.0	Filter press cake from waste water pretreatment
181	37	0.75	37.5	Lab pack outdated corrosive
181	37	0.10	5.0	Spoiled chromic acid mixture
181	39	4.10	205.0	Misc electroplating waste
181	36	24.15	1207.5	Overhauling of plating vault and wastewater system
181	29	90.00	4500.0	Contaminated sulfur from SRU
181	10	10.00	500.0	Filtration
181	28	5.00	250.0	Spill
181	32	1.50	75.0	Clean out of furnace stacks
181	32	340.00	17000.0	Glass mfg.
181	32	59.00	2950.0	Glass mfg.
181	36	5.04	252.0	Electropolish tank sludge
181	36	90.00	4500.0	Waste water treatment system
181	37	0.08	4.0	Nitrogen tetraoxide removed from shuttle fuel/oxidizer system
181	36	0.90	45.0	Waste acid cleaning solution from circuit board cleaning operat.
181	36	0.30	15.0	Waste tin plating solution
181	36	2.50	125.0	Waste from circuit board cleaning operation
181	36	10.00	500.0	Waste copper plating solution
181	36	5.30	265.0	Obsolete solder plating solution
181	36	1.00	50.0	Spent circuit board cleaning solution
181	36	0.40	20.0	Spent circuit board plating solution
181	99	2.85	142.5	Electroplating
181	29	1.00	50.0	Internals from a vacuum distillation column
181	36	2.50	125.0	Printed circuit mfg.
181	28	17.00	850.0	Fly ash cleanout from waste heat boiler
181	36	3.00	150.0	Electroless nickel plating
181	37	7.31	365.5	Anodize seal of metal parts
181	29	148.60	7430.0	Beavon tail gas sulfur recovery process
181	29	2.70	135.0	Cleaning of heat exchanger bundle
181	29	3.50	175.0	Removal of sludge from cooling tower
181	28	3.50	175.0	Spill cleanup material
181	39	0.30	15.0	Stripping process sodium cyanide
181	28	200.00	10000.0	Ammonium nitrate reclamation
181	34	10.00	500.0	Zinc phosphate bath
181	29	10.00	500.0	Cleaning of lime slurry strainer baskets
211	36	0.83	41.5	Cleaning in semiconductor processing
211	51	583.33	29166.5	Contaminated rainwater
211	51	3.00	150.0	Empty sample bottles crushed
211	49	0.20	10.0	Lab sample preparation
212	28	162.20	8110.0	Under water cleanup
212	38	0.62	31.0	Cleaning optical components
212	39	0.23	11.5	Empty drums: last contained flammable material
213	28	83.33	4166.5	Produced when washing resin kettle between batches
213	37	1.00	50.0	Water contaminated gas
221	28	10.42	521.0	Site cleanup material
221	37	23.33	1166.5	90 day clarifier clean according to SQAMD requirements
221	97	41.67	2083.5	Slop collection tank

(continued)

Table 5.13. (continued)

221	10	87.50	4375.0	Flushing of gas storage tank prior to removal
221	37	9.00	450.0	Cleaning oil tanks
221	48	0.10	5.0	Pumpings from manhole
221	42	12.30	615.0	Steam cleaner residue
222	34	1.00	50.0	Oil quench bath residue
222	35	41.00	2050.0	Clarifier for ind. waste water
222	29	70.00	3500.0	Cleaning of API separator
222	37	1650.00	82500.0	Removing oil/water ballast from ocean vessels
222	28	66.67	3333.5	Lube oil additive manufacture
222	25	21.00	1050.0	Site cleanup material
222	29	129.17	6458.5	Crude & fuel oil dehydration
222	28	14.00	700.0	Manufacturing of dispersions
222	39	4.20	210.0	Underground fuel tank
222	34	20.83	1041.5	Clarifier rinsewater
222	95	8.00	400.0	Mechanical repair shop operations
222	24	71.25	3562.5	Removal of discarded fuel oil tank
222	28	27.00	1350.0	Rain water run off into containment area
222	73	6.67	333.5	Residue from cleanup of empty drums
223	37	148.49	7424.5	Machining of rocket engine parts
223	13	25.00	1250.0	Oil well maintenance
223	37	1850.00	92500.0	Cleaning of barges, tugs and oil spill containment boom
223	37	0.60	30.0	Normal oil spillage in maintenance yard
223	37	10.00	500.0	Lube oil-contaminated water from metal forming machines
223	48	3.00	150.0	Pumpings from manholes
223	48	20.90	1045.0	Pumpings from manholes
223	48	13.00	650.0	Pumping from manholes
223	49	12.50	625.0	Steam cleaning waste from cleaning a gasoline truck
223	29	53.70	2685.0	Misc. oily sludge
223	29	362.00	18100.0	Cleanup material from wastewater treatment facilities
223	34	6.70	335.0	Cleaning of metal parts
231	28	50.00	2500.0	Rinsate from agricultural spraying equip.
231	07	8.12	406.0	Pesticide rinse
231	28	7104.00	355200.0	Pesticide formulation waste water
231	95	8.33	416.5	Rinse water for cleanup in rodenticide formulation process
231	28	3237.50	161875.0	Equipment cleaning
231	07	62.50	3125.0	Excess agricultural chemical sprays
231	7	114.00	5700.0	Pesticide application
231	07	41.67	2083.5	Pesticide application equip. rinsing
231	07	20.83	1041.5	Pesticide application/transportation
231	07	104.17	5208.5	Pesticide application/transportation
231	07	6.00	300.0	Pesticide application
231	07	41.67	2083.5	Pesticide application
232	28	2252.00	112600.0	Manufacture of 10,10-oxybisphenoxarsine
232	28	1.00	50.0	Cleanout iron bearing sludge from storage tank & solidification
232	28	23.00	1150.0	From mfg. activity of organic pesticide chemicals
232	07	3.30	165.0	Site cleanup material
241	13	20.00	1000.0	Crude oil separation and sludge
241	20	4.60	230.0	Spill/site clean up material containing

(continued)

Table 5.13. (continued)

				zinc chromate
241	29	0.32	16.0	Chemical cleaning
241	48	10.42	521.0	Wash out of tank before removal
241	10	10.42	521.0	Flushing inactive gasoline storage tank prior to destruction
241	29	1060.00	53000.0	Cleaning wastewater tanks
241	97	5.00	250.0	Clean up of fuel spill from refueling equip.
241	29	67.40	3370.0	Tank bottom waste
241	29	3970.00	198500.0	Tank cleaning
241	99	2.62	131.0	Removal of 1000 gallon gasoline storage
241	29	29.13	1456.5	Equipment cleanup material
241	29	36.20	1810.0	Equipment cleanup material
241	29	136.10	6805.0	Equipment cleanup material
241	29	86.19	4309.5	Equipment cleanup material
241	29	13.50	675.0	Cleaning of sludge from tankage
241	29	15.00	750.0	Cleaning of sediment from tankage
241	28	73.00	3650.0	Tank bottom sediment
241	28	2.00	100.0	Tank bottom sediment
241	29	48.50	2425.0	Site cleanup/tank replacement
241	29	3.20	160.0	Tank replacement/site cleanup
241	29	5.70	285.0	Tank cleaning
241	29	44.30	2215.0	Contaminated soil
241	29	40.00	2000.0	Site cleanup/tank replacement
241	29	3.00	150.0	Site cleanup/tank replacement
241	29	77.80	3890.0	Site cleanup/tank replacement
241	34	741.67	37083.5	Tank bottoms and spent solution
251	34	1.00	50.0	Degreasing metal parts
252	29	10.00	500.0	Distillation bottoms from sulfinol reclaimer
261	14	0.56	28.0	Capacitor removal
261	24	1.35	67.5	Used capacitors (sawmill & plywood mfg.)
261	95	1.00	50.0	Transformers and site cleanup material
261	37	0.21	10.5	Capacitors removed from locomotive alternators to be rebuilt
261	29	0.50	25.0	Electrical equipment (transformers, capacitors)
261	28	22.80	1140.0	Transformers and capacitors
261	49	104.00	5200.0	Drained transformers, capacitors
261	49	2.00	100.0	Drained transformers, capacitors
261	28	34.00	1700.0	Site cleanup - removal of underground vessel
261	97	0.05	2.5	Electrical equipment replacement
261	97	14.80	740.0	Transformer
261	49	68.40	3420.0	Disposal of used transformer oil & transformers
261	49	1.00	50.0	Site cleanup of material from electrical equipment failure
261	37	19.00	950.0	Fluorescent light ballasts
261	37	0.30	15.0	Replacement of switchbox & cable
261	37	16.50	825.0	Removal of PCB transformer
261	99	202.40	10120.0	Waste transformers containing PCB
261	73	0.62	31.0	Removal of PCBs
261	13	1.00	50.0	Site cleanup material
261	13	4.10	205.0	Oil field waste
261	13	31.00	1550.0	Disposal of electrical equipment
261	39	1.40	70.0	Fluorescent light ballast replaced
271	39	0.40	20.0	Product coating

(continued)

Table 5.13. (continued)

271	28	120.00	6000.0	Cleanup water from water based primer paint line
	28	150.00	7500.0	Paint manufacture
291	95	0.83	41.5	Damaged through warehouse handling
321	28	60.00.00	3000.0	Mfg. of water base adhesives
321	37	97.00	4850.0	Chrome destruct system
321	97	700.00	35000.0	Sewage treatment plant
322	29	438.00	21900.0	Wastewater treatment plant biological waste
	99	0.35	17.5	Circuit board fabrication
331	38	0.05	2.5	Manufacturing
331	39	0.21	10.5	Raw material
331	28	6.65	332.5	Off-spec and obsolete material for disposal
341	48	4.17	208.5	Tank bottom removal
342	28	4.17	208.5	Testing polymer blending equipment
342	36	17.50	875.0	Dump liquid
43	13	2873.00	143650.0	Oil well production
43	28	28.00	1400.0	Chemical mfg.
43	49	3.20	160.0	Drain cooling system
343	32	0.42	21.0	Disposal of obsolete water treatment materials
343	48	25.00	1250.0	Pumpings from manholes
343	48	6.00	300.0	Pumpings from manholes
343	49	92.40	4620.0	Flushing contaminated tank
343	49	33.20	1660.0	Tank flushing
343	49	10.40	520.0	Tank flushing
343	37	20.83	1041.5	Spill cleanup
343	13	83.00	4150.0	Reboiler concentrating glycol at gas plant
351	36	11.10	555.0	Spill
352	34	0.75	37.5	Painting metal parts
352	36	15.60	780.0	Electroplating
352	26	0.75	37.5	Chemical and pharmaceutical mfg.
352	97	4.00	200.0	Aircraft/vehicle maintenance oil spill cleanup
352	37	2.63	131.5	Paint stripper tank sludge, stripping of paint shop screens
352	37	0.50	25.0	Urethane foam for packing & crating equip.
352	37	9.30	465.0	Underground tank excavation
352	39	37.50	1875.0	Cleanup
352	82	8.48	424.0	Spill cleanup
411	34	113.00	5650.0	Steel cleaning
411	28	1100.00	55000.0	Aluminum sulfate mfg.
411	36	94.41	4720.5	Treatment of process waste water
421	36	205.00	10250.0	Sludge
421	36	70.00	3500.0	Process water treatment
421	38	169.00	8450.0	Waste treatment
421	36	1228.00	61400.0	Waste treatment sludge
421	97	158.40	7920.0	Industrial waste plant
421	37	101.00	5050.0	Waste water treatment
421	34	150.00	7500.0	Filter press cake generated by wastewater pretreat. system
421	39	92.50	4625.0	Filter press cake from electroplating waste treatment
421	49	200.00	10000.0	Lime precipitation
421	49	100.00	5000.0	Lime precipitation
421	49	30.00	1500.0	Lime precipitation
421	49	70.00	3500.0	Lime precipitation

(continued)

Table 5.13. (continued)

421	36	9.00	450.0	Precipitation of heavy metals from waste water
421	37	1445.20	72260.0	Electroplating operation
431	37	140.60	7030.0	Cleaning out bonderite system
431	34	110.00	5500.0	Tank bottom sludge from ppt. of zinc phosphate coating of coils
431	34	2.00	100.0	Rust inhibitor (dipping operation)
441	49	11602.00	580100.0	Hydrogen sulfide abatement systems in geothermal power plts.
441	49	7857.00	392850.0	Hydrogen sulfide abatement sys. in geothermal power plants
441	29	932.30	46615.0	Sulfur solids from Stretford unit
451	29	200.00	10000.0	Equipment cleaning
461	28	50.00	2500.0	Crushed fiber drums
461	25	111.00	5550.0	Site cleanup material
461	28	6.00	300.0	Tub cleaning
481	29	64.66	3233.0	Tank clean-out
491	28	60.00	3000.0	Wastewater filtration
491	34	1.04	52.0	Site cleanup material
491	27	3.75	187.5	Bulk storage tank cleaning
491	29	40.00	2000.0	Cleaning of cooling tower basin
491	29	20.00	1000.0	Cooling tower basin sludge
491	37	215.00	10750.0	Neutralization of titanium etchants and rinsewater
491	28	160.00	8000.0	Manufacturing of zinc chloride
491	34	11.20	560.0	Interceptor waste water and sludge
491	29	150.00	7500.0	Petroleum refining
491	36	0.33	16.5	Manufacture of indium phosphide crystals
491	39	109.98	5499.0	Wastewater treatment sludge from plating bath rinse water
491	29	19.00	950.0	Petroleum refining
491	28	7.51	375.5	Sludge & water from clarifier
491	89	0.30	15.0	Spill cleanup
491	99	16.67	833.5	Sludge from holding tanks
491	36	14.90	745.0	Semiconductor mfg.
491	36	3.00	150.0	Waste treatment plant sludge from plating operations
491	36	65.40	3270.0	Lapping of quartz crystals, lenses & silicon wafers
491	36	126.50	6325.0	Industrial waste pretreatment
491	36	28.20	1410.0	Deburring, phosphate process rinse, plastic material stripping
491	99	14.58	729.0	Clarifier sludge
491	27	5.00	250.0	Waste water and sludge from a sump connected to sewer system
491	28	5.00	250.0	Mfg. of photographic & electronic chemicals
491	36	57.70	2885.0	Waste water pretreatment
491	34	5.00	250.0	Aluminum anodizing etching processes
491	35	16.02	801.0	Waste collection resulting from the process of belt grinding, etc.
491	36	13.75	687.5	Heavy metal hydroxide precipitation
491	29	15.40	770.0	Maintenance heavy oil processing
491	36	121.00	6050.0	Electroplating circuit boards
491	36	1.58	79.0	Clean out of acid fume scrubber

(continued)

Table 5.13. (continued)

491	28	162.50	8125.0	Closure of industrial waste surface impoundment settling pond
491	36	12.00	600.0	Clarifier sludge
491	37	154.17	7708.5	Paint operations
491	36	4.80	240.0	Neutralization of metal finishing solution
491	29	39.30	1965.0	Cleaning sludge from cooling tower
491	36	0.08	4.0	Spill cleanup
491	36	29.00	1450.0	Sludge precipitate from rinse water
491	29	66.30	3315.0	Cleanout of cooling tower
491	28	1000.00	50000.0	Wastewater treatment system
491	33	75.00	3750.0	Caustic cleaning of steel strip
491	29	250.00	12500.0	Petroleum refining - cooling tower blowdown
491	29	173.00	8650.0	Cleaning of salt water cooling boxes - silt
491	28	48.00	2400.0	Solids that settle out in the industrial waste water system
491	36	12.00	600.0	Treatment plant sludge
491	36	55.00	2750.0	Treatment plant sludge
491	34	63.00	3150.0	Wastewater pretreat
511	51	271.20	13560.0	Wholesale trade
512	37	5.05	252.5	Propellant production
512	36	1.00	50.0	Semiconductor processing
512	39	0.13	6.5	Drums last containing flammable solvents
512	28	12.00	600.0	Empty polyethylene liners with raw material residue
512	37	82.00	4100.0	Painting operations crushed contaminated drums
512	28	0.50	25.0	Drums which contained unusable product
512	36	2.00	100.0	Printed circuit board mfg.
512	28	22.00	1100.0	Underground tank excavation
512	37	2.00	100.0	Crushed, empty coolant drums
512	36	3.78	189.0	Removal of old plating tanks
512	28	50.00	2500.0	Mfg. of specialty chemicals (empty containers)
512	49	122.00	6100.0	Disposal of drums previously containing hazardous materials
512	49	1.40	70.0	Removal of underground tank - last contained gasoline.
512	39	0.25	12.5	Empty drums from facility operations
512	29	59.60	2980.0	Disposal of crushed drums
512	36	0.03	1.5	Empty freon containers
512	51	271.20	13560.0	Wholesale trade
512	36	0.05	2.5	Chemical laboratory
512	36	0.25	12.5	Empty containers from plating shop
513	28	31.50	1575.0	Site cleanup material
513	28	1.60	80.0	Synthesis & formulation of pressure-sensitive adhesives
513	35	3.00	150.0	Touchup paint
513	29	50.00	2500.0	Offspec. empty containers (crushed)
513	36	185.00	9250.0	Empty containers-flammables
513	28	0.50	25.0	Lab supplies
513	37	10.00	500.0	Chemicals for printed circuit board mfg & aircraft parts
513	36	25.00	1250.0	Semiconductor mfg.
513	07	9.00	450.0	Aerial application of pesticides
513	36	0.40	20.0	Degreasing buckets
513	07	7.48	374.0	Pesticide and fertilizer application

(continued)

Table 5.13. (continued)

521	13	1429.00	71450.0	Oil well drilling
521	13	53.00	2650.0	Crude oil production
521	13	1500.00	75000.0	Oil and gas production
521	36	568.00	28400.0	Monitoring well installation
521	13	1.40	70.0	Drilling
521	13	74.00	3700.0	Oil production/oil well drilling waste
521	13	20.40	1020.0	Oil production
521	13	1444.00	72200.0	Drilling oil and gas wells
531	97	2841.67	142083.5	Chemical toilet cleanout
541	59	3.33	166.5	Retail trade
541	59	8.75	437.5	Waste photochemicals
541	59	1.52	76.0	Photoprocessing
541	72	15.02	751.0	Waste photochemicals
541	59	10.00	500.0	Waste photochemical
541	37	20.00	1000.0	Photographic solutions
541	99	89.00	4450.0	Substance used for photographic processing
541	37	39.00	1950.0	Photoprocessing
541	59	0.62	31.0	Photo development
541	59	1.25	62.5	Photo development
541	59	15.00	750.0	Photo development
541	59	10.00	500.0	Waste photochemical
541	39	84.00	4200.0	Photoprocessing
541	39	350.00	17500.0	Photoprocessing
541	59	1.00	50.0	Waste photochemicals
541	59	10.00	500.0	Photo development
541	36	4.17	208.5	Fixer from photo labs
541	59	0.60	30.0	Photoprocessing
541	99	3.12	156.0	Waste photochemicals
541	99	3.12	156.0	Waste photochemicals
541	37	0.10	5.0	On-site photo lab
541	59	7.50	375.0	Waste photochemical
541	99	3.12	156.0	Waste photochemicals
551	36	0.05	2.5	Empty freon containers
551	28	4.30	215.0	Laboratory waste chemicals
551	38	0.04	2.0	Chemical laboratory waste
551	36	0.69	34.5	Semiconductor R&D
551	28	0.01	0.5	Off spec chemical discovered during stockroom cleanout
551	99	1.30	65.0	Spill cleanup
551	36	0.05	2.5	Spent or outdated chemicals used in mfg of semiconductors
551	39	0.35	17.5	R&D lab waste
551	28	0.25	12.5	Manufacture of photographic & electronic chemicals
551	99	32.00	1600.0	Waste liq.from pilot plant exper. runs on proprietary pesticide
551	36	0.30	15.0	Painting waste
551	37	104.00	5200.0	Lab waste
551	36	2.20	110.0	Lab chemicals past shelf life time limits
557	36	0.69	34.5	Semiconductor R&D
561	28	40.00	2000.0	Wash out sufactant mixing equipment
561	99	10.50	525.0	Wash. commercial linens/uniforms w/detergents, soaps
571	28	6.00	300.0	Combustion of spent sulfuric acid

(continued)

Table 5.13. (continued)

571	95	680.00	34,000.0	Sewage sludge incineration
571	49	113.00	5,650.0	Boiler and associated equip. cleaning
571	29	169.40	8,470.0	Removal of fly ash, refractory & slag from furnaces
571	29	13.30	665.0	Sandblasting of boiler firebox
571	33	600.00	30,000.0	Emission control dust from iron foundry cupola furnace
571	28	115.00	5,750.0	Combustion of oil refinery sulfur feedstreams
581	36	4.00	200.0	Used chemical sink ductwork
581	13	2,632.54	131,627.0	Gas scrubbing
581	28	70.00	3500.0	CO_2 plant scrubber - potassium permanganate
591	99	3.00	150.0	Flare testing
591	33	240.00	12,000.0	Grinding dust - titanium
591	34	2.00	100.0	Hot dip galvanizing operation
591	32	500.00	25,000.0	Glass mfg.
611	28	4.00	200.0	Site cleanup material
611	36	133.70	6,685.0	Cleanup of soil and ground water
611	13	100.00	5,000.0	Chemical spills
711	34	33.33	1,666.5	Precious metal finishing process containing cyanides
711	99	5.00	250.0	Plating rinses/spent stripping solution
711	37	6.00	300.0	Cadmium plating solution
711	99	13.33	666.5	Silver recovery
711	34	20.83	1041.5	Precious metals finishing process containing cyanides
711	34	2.08	104.0	Zinc plating
711	99	0.62	31.0	Metal cleaning/stripping
711	99	5.00	250.0	Cadmium electroplating solution
721	28	2,200.00	110,000.0	Mfg of freon compounds
721	36	105.07	5253.5	Mfg semiconductors & related devices, calcium fluoride sludge
722	10	3.12	156.0	Hydrometallurgical leach solution
723	34	5.83	291.5	Spent chromic acid anodize solution (plating solution)
723	34	6.25	312.5	Chromic acid anodizing solution
723	36	83.33	4166.5	Etching in electroplating operation
723	34	11.25	562.5	Waste water rinsing and cleaning sludges from electroplating
723	34	18.00	900.0	Chrome coating of aluminum prior to painting
723	28	14.00	700.0	Collection of water used for tank cleaning
723	99	125.00	6250.0	Plating rinse
723	37	55.00	2750.0	Aluminum alkaline cleaner
723	37	380.00	19000.0	Chromic acid dragout pH 0-3
723	37	170.00	8500.0	Sulfuric acid anodize seal solution pH 5-6
723	37	50.00	2500.0	Aluminum oxidizer process tank
723	34	2.50	125.0	Metal services electroplating
723	39	27.92	1396.0	Spent chromium liquid used in plating process
723	28	23.00	1150.0	Electroplating
723	37	18.00	900.0	Process solution for sealing anodized aluminum
723	34	27.08	1,354.0	Electroplating operations
723	34	30.00	1,500.0	Chrome stripping and nickel plating
723	36	25.00	1,250.0	Semiconductor mfg.
723	34	26.00	1,300.0	Chromium plating waste

(continued)

Table 5.13. (continued)

723	49	14.00	700.0	Engine jacket water with chromate inhibitor
723	99	73.80	3690.0	Chromate conversion coating process
723	36	2.80	140.0	Deoxidizing aluminum & etching copper
723	33	4,500.00	225,000.0	Pickling & chemical treatment of steel (electroplating)
724	36	33.33	1,666.5	Stripping lead in electroplating operation
724	36	17.19	859.5	Solder plating solution
724	32	314.54	15,727.0	Waste tile glaze from ceramic tile prod/solids from filter press
724	39	0.30	15.0	Plating waste
726	34	2.92	146.0	Metal finishing
726	34	175.00	8,750.0	Spent electroless nickel plating solution
26	34	4.17	208.5	Precious metal plating on electrical contacts
26	28	0.69	34.5	Nickel plating solution
726	34	8.33	416.5	Electroplating
736	34	113.00	5,650.0	Spent electroless nickel plating solution
751	29	0.48	24.0	Chemical cleaning
751	34	1.50	75.0	Degreaser
791	36	1,166.67	58,333.5	Etching process in semiconductor production

5.5.1. Selection of Wastestream Suitable for Chemical Fixation

A wastestream was chosen by selecting a wastestream from the BGR database. The wastestream was checked to determine if the components contained asbestos, organics or cyanides. If the components contained asbestos, chemical fixation was not considered appropriate. If the components contained organic compounds or cyanides, their concentrations within the stream were summed and the results were compared with the treatability limits of 1% for organic compounds or 0.3% for cyanides. Any wastestream that contained more than 1% organic compounds or 0.3% cyanides was excluded from consideration for chemical fixation.

After checking the components for asbestos, organic compounds and cyanides, we determined if the components contained any of the metals as listed in Table 5.10. Once the component was found to contain one or more of the "treatable" metals, the individual metal concentration of this component in the wastestream was calculated. This procedure was repeated for the rest of components in the stream. The contributions from all metals that contribute to the treatable components in the waste were summed. If the concentrations of the

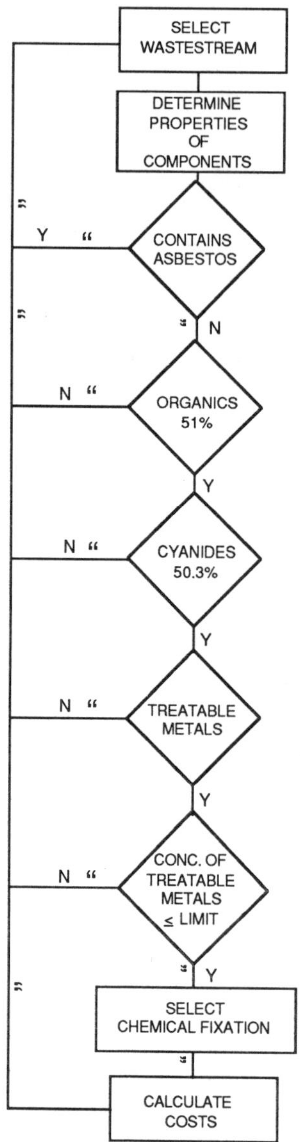

Figure 5.1 Flow Diagram for Selection and Cost Estimation of Chemical Fixation Process

treatable metals in the waste were all less than the limits set forth in Table 5.10, chemical fixation was deemed appropriate for treatment. Even if only one of the metals in the waste failed to meet these limits, the wastestream was not considered suitable for chemical fixation. Once a wastestream was selected for chemical fixation, the cost of treating the stream was calculated.

Since the treatability of components and their concentration limits for selection of fixation were adopted from those appropriate to the CHEMFIX process, caution must be exercised when analyzing the results of this study. As discussed in Section 5.4, other chemical fixation processes, due to their proprietary additives, may be able to treat some components which cannot be handled by the CHEMFIX process. Our results will be process-specific although we believe that they are quite reliable because of the background information available on the CHEMFIX process.

5.5.2. Cost Estimation for Chemical Fixation

Since the criteria for chemical fixation were based on the CHEMFIX process, cost data for the CHEMFIX process were used in the estimation of cost of treatment. The price range of the CHEMFIX process is \$25-75 per ton of waste; therefore the average unit price (\$50/ton) was chosen to calculate the fixation cost. This price included equipment set-up on-site, operating personnel and reagent costs. No capital costs were reported from the vendor; however, it is believed that the capital cost is small since it consists only of the cost of mixers, hoppers and conveyers. Also, it might be possible to rent the equipment, thereby avoiding purchasing equipment which might not be used subsequently. Finally, we calculated the cost using:

$$\text{Cost of fixation} = 50 \times \text{Annual Volume (tons)} \tag{5.1}$$

5.5.3. Results on Treatability Using Chemical Fixation

The results of the program are summarized in Table 5.13, which contains data on wastestreams treatable by chemical fixation. Approximately one-eighth of the wastestreams can be treated by chemical fixation. Among the 1221 treatable wastestreams, 368 contain cyanides or organic compounds. Thus, about 70% of the treatable wastestreams containing metals did not contain any cyanides or organic compounds.

Most of the treatable wastestreams originate in the electroplating or metal industry, as evidenced by the descriptions of the processes from which they were generated. To further substantiate this conclusions, Tables 5.14 and 5.15 list the top ten California Waste Category (CAL) Codes and SIC Codes, respectively, of the treatable wastestreams. Six of the top ten CAL Codes from Table 5.14 (111, Acid Solution with Metals; 541, Photochemicals/photoprocessing waste; 132, Aqueous Solution with Metals; 181, Other Inorganic Solid Waste; 171, Metal Sludge; and 121, Alkaline Solution with Metals) contain treatable metals (Table 5.12). The total number of wastestreams in those six CAL Codes accounted for 47% of the treatable wastestreams. About 41% of the treatable wastestreams were from the first three SIC industry groups as shown in Table 5.15. Those three SIC groups are: 36 (Electrical and Electronic Machinery), 34 (Fabricated Metal Products) and 37 (Transportation).

Table 5.14. Top 10 CAL Groups of California Wastes Selected for Chemical Fixation

CAL Code	# of Wastestreams
111	205
541	112
132	98
181	76
491	55
241	44
171	43
135	41
121	37
512	35

Although chemical fixation is suitable for the treatment of metal-containing wastes, it may not be advantageous, from the economic point of view, to treat high-volume (aqueous) wastes with very low metal concentrations in this fashion. This is supported by a few wastestreams listed in Table 5.13, for which the costs of treatment were in the millions of dollars due to the high volume of waste. Under such conditions, it may be more economical to use other treatment methods, such as, ion-exchange adsorption.

Table 5.15. Top 10 SIC Groups of California Wastes Selected for Chemical Fixation

SIC Code	# of Wastestreams
36	205
34	158
37	141
28	136
99	98
29	96
39	79
59	47
49	35
13	39

5.6. REMARKS ON THE USEFULNESS AND APPLICABILITY OF THE TREATMENT ESTIMATION METHODOLOGY

The techniques we have used to estimate the applicability of chemical fixation to treat a wastestream require two types of information:

1. Components and their concentrations - These are used to determine the feasibility of chemical fixation. If one or more components in the wastestream contained the metals treatable by

chemical fixation (Table 5.12), and their concentrations were within the treatability limits, chemical fixation was deemed suitable for that wastestream.

2. Annual volume of the wastestream - These are used to determine the cost of fixation.

In order to calculate the fixation cost more accurately, it would be necessary to know other characteristics of the wastestream, such as its pH. Chemical fixation is best suited to wastestreams having high pH values, and it is significantly affected by a drop in pH. If the pH of the wastestream is known, the amount of alkali needed to adjust the pH can be determined exactly. This would provide a better estimate of the cost of reagent (although it is a small part of the cost) in the calculation of the fixation cost.

5.7. REFERENCES

1. Poon, C. S., C.J. Peters and R. Perry, "Use of Stabilization Processes in the Control of Toxic Wastes," Effluent and Water Treatment Journal, p. 451, Nov., 1983.

2. Wiles, C.C., "Perspectives on Solidification/Stabilization Technology for Testing Hazardous Waste," EPA/600/D-87/027, January, 1987.

3. Conner, J.R., "Fixation and Solidification of Wastes," Chemical Engineering, p. 79, Nov. 10, 1986.

4. Potter, J., ed., "Stabilization Technologies," Alternative Technology for Recycling and Treatment of Hazardous Wastes, California Dept. of Health Services, July 1986.

5. Riemann, P., "Solidification of Inorganic Wastes." 2nd. International Symposium on Operating European Centralized Hazardous (Chemical) Waste Management Facilities, Denmark, Sept. 1984.

6. "Survey of Solidification/Stabilization Technology for Hazardous Industrial Wastes," EPA-600/2-79-056, July, 1979.

7. Tittlebaum, M.E., Seals, R.K., Cartledge, F.K., and Engels, S., "State of the Art on Stabilization of Hazardous Organic Liquid Wastes and Sludges."CRC Critical Reviews in Environment Control, volume 15, issue 2, 1985.

8. Bogue, R.H., The Chemistry of Portland Cement, 2nd ed., Reinhold, New York, 793 pp., 1955.

9. American Society for Testing and Materials, Annual Book of ASTM Standards, Part II, Philadelphia, PA, 1973.

10. Department of the Army, Engineering and Design - Laboratory Soil Testing, Engineering Manual, EM 1110-2-1906. Washington D.C., Nov. 30, 1970.

11. EP Toxicity Test Procedure, 40 CFR 261.24, U.S. Federal Register, May 19, 1980.

12. Toxic Characteristic Leaching Procedure, U.S. Federal Register, V. 51, No. 114, June 13, 1986.

13. Multiple Extraction Procedure Methodology, U.S. Federal Register, V. 47, No. 225, Nov. 22, 1982.

14. Bishop, P. and D. Gress, "Leaching from Stabilized/Solidified Hazardous Wastes," Proceedings of National Conference on Environment Engineering, Minneapolis, MN, 1982.

15. Escher, E.D., and J.W. Newton, "The Determination of Fixation Treatment Method Limits for Hazardous Liquids and Industrial Sludges from Disparate Sources," NUS Corporation, Pittsburgh, PA 15275.

16. Technical information from Chemfix Technologies, Inc., Metairie, LA 70001.

17. Technical information from Westinghouse Electric Corp., Madison, PA 15663-0286.

18. Technical information from Chemical Waste Management, Oak Brook, IL 60521.

19. Technical information from Mittelhauser Corp., Laguna Hills, CA 92653.

20. Technical information from Ensotech, Inc., North Hollywood, CA 91605.

21. R.L. Bell, A.P. Jackman and R.L. Powell, a report submitted to the California Dept. of Health Services by Dept. of Chemical Engr., U. of California, Davis, 1988.

6. Chemical Oxidation

6.1. INTRODUCTION

Chemical oxidation has been found very effective in the treatment of industrial and domestic wastewater. In particular, oxidation offers one of the few methods for removing odor, color, and various potentially toxic organic substances like phenols, pesticides and industrial solvents [1,2]. It also disinfects drinking water by killing or inactivating pathogenic microorganisms that may be present [3]. In this section we discuss the usefulness of chemical oxidation for the destruction of hazardous wastes in aqueous solutions. Three types of chemical oxidants are described: chlorine, chlorine dioxide, and ozone. Attendant technologies with flowsheets and implementation costs are given. Compounds which can be treated by these technologies are listed along with the typical levels of destruction which can be obtained.

6.2. DESCRIPTIONS OF CHEMICAL OXIDATION TREATMENT PROCESSES

6.2.1. Chlorination

Chlorine is the most widely used chemical oxidant for disinfecting drinking water [4]. Its use started in Europe in the late 19th century, and later spread to North America. Since it is inexpensive to implement and it is an available proven technology, it has found widespread use. One of its most serious drawbacks is that chemical oxidation by chlorination can produce toxic trihalomethanes when organics are present in the aqueous phase [1].

Chlorine is a strong oxidizing agent. When dissolved in water, chlorine is hydrated to form chloride, Cl^-, and hypochlorite, ClO^- ions [1]. The reactions are:

$$Cl_2 + H_2O \longrightarrow HClO + H^+ + Cl^- \quad (6.1)$$

$$HClO \longrightarrow H^+ + ClO^- \quad (6.2)$$

The oxidizing ability of these two ions is responsible for oxidation by chlorination. Chlorine oxidation is very effective for removing cyanide from wastes produced by the electroplating or metal finishing industries [1]. It can be implemented on a small scale, which is appropriate for the industrial units typical of electroplaters and metal finishers. Processing such wastes has increased greatly in recent years. This is due to increased regulation of the levels of cyanides in wastewaters. For example, since June 1, 1983, liquid wastes containing more than 1000 mg/l of free cyanides have been banned from land disposal in California [5]. This has led many producers of such wastes to pretreat their cyanide containing aqueous wastestreams by chlorination. Properly treated wastes can then be released to the sewer. Besides cyanides, chlorine also attacks certain organic compounds in wastewater such as phenols, acids and alcohols. Typical technologies for treating cyanide containing wastestreams, and other aqueous wastestreams are described in the next Section.

6.2.1.a. Process Descriptions

Treatment of Cyanide Containing Wastes

The oxidation of cyanides in wastewater by chlorine has been the most widely acceptable method of cyanide treatment for the past thirty years [5]. In an alkaline solution of pH 8.5 or higher, chlorine reacts with cyanide to form cyanate [5,6]. The reaction is described by:

$$CN^- + Cl_2 + 2OH^- \longrightarrow CNO^- + 2Cl^- + H_2O \qquad (6.3)$$

This reaction requires 6.8 pounds of chlorine to convert one pound of cyanide to cyanate [7]. Usually more than 6.8 pounds of chlorine is used due to the presence of other oxidizable components. If the pH is controlled to between 8.5 and 9.0, at ambient temperature and pressure it takes ten to thirty minutes for one hundred percent conversion of free cyanide to cyanate. If the pH is increased, the reaction time diminishes. However, if pH drops to as low as 8.0, not only does the conversion decrease tremendously, but also highly toxic cyanogen chloride, CNCl, begins to form. This highly toxic substance (e.g., CNCl is as toxic as hydrogen chloride) is produced via:

$$CN^- + Cl_2 \longrightarrow CNCl + Cl^- \qquad (6.4)$$

It is therefore very important to maintain the pH level above 8.5 when chlorinating cyanides.

The cyanate which is formed by the reaction (6.3) is further converted to bicarbonate and nitrogen at a pH of 8.5 to 9.5. The overall reaction is [4]:

$$3Cl_2 + 2CNO^- + 6OH^- \longrightarrow 2HCO_3^- + N_2 + 6Cl^- + 2H_2O \qquad (6.5)$$

Again, the pH must be maintained above 8.5 to inhibit the formation of toxic by-products, in this case, chloramines. Chlorination reactions are commonly carried out at ambient temperature and pressure. Besides pH, the effectiveness of destruction of cyanide depends on factors such as initial cyanide concentration, presence of metal ions like nickel, iron and cobalt, and mass transfer effects on the chlorine in solution.

Chlorination is typically used to treat dilute cyanide solutions and rinse waters having cyanide concentrations below 1000 mg/l [5]. For spent process solutions (which have cyanide concentrations well above 1000 mg/l), however, the chlorination process becomes quite inefficient [5]. Other forms of treatment, such as wet air oxidation (see Chapter 4), are required to obtain a high percentage of cyanide removal. It is also possible to combine such treatment with chlorination in a two stage process which can treat a wide range of wastes to a relatively high degree of destruction.

The presence of metal ions such as nickel, iron and cobalt in cyanide rinse water interfere with the destruction of cyanides to cyanates. Nickel forms a cyanide complex which cannot be completely transformed to cyanate in a relatively short time by chlorine. It is necessary to remove the nickel ion first by metal ion precipitation and then proceed with a chlorination step to remove all the cyanides.

All chlorination technologies require that the chlorine dissolves in the aqueous phase prior to the oxidation step. The rate of cyanide destruction therefore depends on the dissolution rate of chlorine in aqueous solutions. To minimize mass transfer limitations, a well-designed bubbler or dispersion system is usually used to increase the gas-liquid interfacial transfer area.

A schematic diagram of a two stage chlorination system is shown in Figure 6.1 [8]. Chlorine, which is produced in the chlorinator, is mixed with alkali to raise the pH of the solution. The pH of the solution is maintained at an optimal preset value by the use of a pH meter and an oxidation-reduction-potential (ORP) controller. The liquid effluent from the first reactor (which contains mostly cyanates) then enters the second reactor where it is further converted to nitrogen and carbon dioxide (from bicarbonate decomposition). The reaction time needed in each stage of the process in Figure 6.1 is approximately twenty five to thirty minutes. The degree of conversion of cyanides can be evaluated from samples of the effluent stream by analyzing for the cyanide content using gas chromatography [9] as well as from the dissolved chlorine concentration by the amperometrical method [10] (i.e., by measuring the current in the solution and converting it to concentration using a calibration curve). For many cyanide containing wastestreams, such a process will typically result in an effluent which can be safely discharged to municipal sewers under most existing regulations.

Table 6.1 gives the performance of some alkali chlorination systems [8,11] for influent cyanide concentrations ranging from 0.2 mg/l to 700. The removal efficiencies range from 58% to 100%. It is important to note that the lowest removal efficiencies were obtained at the lowest value of pH, demonstrating the need to process wastes at a high pH in order to obtain the highest removal efficiencies in the shortest time. Most of the data are for iron cyanide, where tests were run over a wide range of operating conditions. The percentage of cyanide removal in Table 6.1 increases slightly with increasing pH, with the optimal pH values being above 10.6. It is generally found that sixty minutes is an optimal contact time. At a fixed chlorine concentration and pH, contacting times both below and above sixty minutes result in decreased destruction of the iron cyanide. Generally,

Table 6.1. Treatment and Removal Levels for Cyanide Waste by Chlorination

Component	Conc. (mg/l) Initial	Final	Percent Removal	Chlorine Conc. (mg/l)	Contact Time, min.	pH
Cyanide	700.0	0.0	100	-	-	-
Cyanide	32.5	0.0	100	-	-	-
Cyanide	5.1	0.1	99	-	-	-
Hydrogen cyn.	6.8	0.0	100	-	-	-
Iron cyanide	0.19	0.08	58	20	30	8.8
Iron cyanide	0.19	0.05	74	20	60	8.8
Iron cyanide	0.19	0.07	63	20	90	8.8
Iron cyanide	0.19	0.04	79	10	30	10.6
Iron cyanide	0.19	0.03	84	10	60	10.6
Iron cyanide	0.19	0.04	79	10	90	10.6
Iron cyanide	0.19	0.03	84	20	30	10.6
Iron cyanide	0.19	0.02	89	20	60	10.6
Iron cyanide	0.19	0.02	89	20	90	10.6
Iron cyanide	0.19	0.03	84	10	30	11.0
Iron cyanide	0.19	0.03	84	10	60	11.0
Iron cyanide	0.19	0.03	84	10	90	11.0
Iron cyanide	0.19	0.01	95	20	30	11.0
Iron cyanide	0.19	0.02	89	20	60	11.0
Iron cyanide	0.19	0.02	89	20	90	11.0

Source: [8,11]

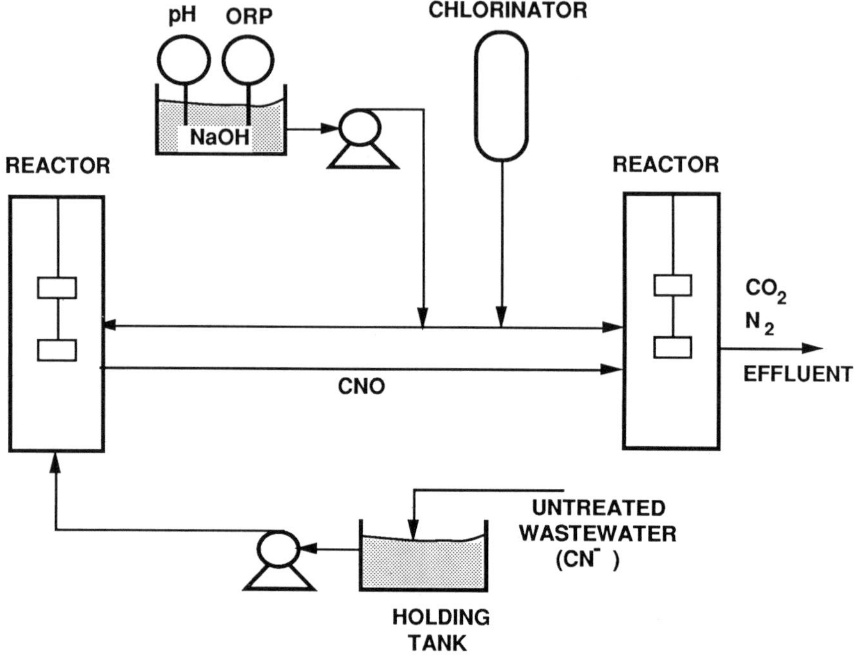

Figure 6.1. Two Stage Chlorination System for Cyanide Removal

destruction increases with the chlorine concentration although this effect diminishes somewhat at higher values of pH. Furthermore, chlorination of cyanide wastes can achieve 90% efficiency (on the average) with concentrations of chlorine lying between 10 and 20 mg/l. On the other hand, there are no reported data for cyanide concentrations above 700 mg/l. This does not imply that chlorination cannot be used to treat concentrations of cyanide wastes higher than 700 mg/l. It likely means, however, that to achieve the same removal percentage in the case of a high influent concentration of cyanide, the process parameters may need to be adjusted (e.g., increasing the residence time or chlorine concentration in the stream, or using a batch/recycle reactor instead of a continuous reactor). Thus, there is no definite cyanide concentration limit for feasibility of chlorination.

For the purpose of the model calculations, we have arbitrarily chosen three concentration limits (0.1%, 1% and 5%) which reflect the degree of contaminant removal (0.1% = high efficiency, 1% = medium efficiency and 5% = low efficiency) by chlorination. That is, based upon the information available to us, we assume that wastestreams containing 0.1% or less of a cyanide contaminant can be very effectively treated using chlorination. Streams containing 5% or more cyanide can be treated, but much less effectively. Those in - between 0.1% and 5% can also be treated, but only to a moderate level of effectiveness. These limits are also applied to chlorine dioxide oxidation and ozonation, which will be discussed in later sections of this chapter.

Some experiments have been done on chlorination of cyanides under ultraviolet radiation. The results indicate that the presence of ultraviolet radiation neither increases nor decreases the percent of removal [10]. As discussed below, this is in marked contrast to many other compounds for which ultraviolet - assisted chlorination is much more effective than just chlorination itself (see Table 6.2).

Treatment of cyanides by chlorination offers the cheapest possible treatment of cyanide containing aqueous wastes from the electroplating and metal finishing industries, provided that the wastestreams contain only a small amount of organics. For high concentrations

Table 6.2. Comparison of Chlorine Consumption with and without Ultraviolet Radiation

Compound	Conc. (ppm)	Chlorine Consumption Irradiated (gg^{-1})*	Not Irradiated (gg^{-1})*
Formic acid	100	0.13	0
Propionic acid	89	0.05	0
Benzoic acid	100	0.40	0
n-Butyric acid	130	0.09	0
Oxalic acid	100	0.45	0
Glutaric acid	132	0.06	0
Adipic acid	102	0.12	0
Fumaric acid	100	0.15	0
Glycolic acid	95	0.11	0
DL-Lactic acid	130	0.20	0
D-Tartaric acid	100	0.14	0
N-Acetyl-L-glutamic acid	96	0.04	0
Formaldehyde	100	0.53	0
Acetaldehyde	98	1.00	0
Methyl ethyl ketone	100	0.18	0
Cyclohexanone	102	0.26	0
Methyl alcohol	100	0.42	0
Ethyl alcohol	100	0.90	0
Isopropyl alcohol	99	0.84	0
n-Butyl alcohol	100	0.46	0
Ethylene glycol	100	0.24	0
Glycerol	100	0.28	0
Sorbitol	100	0.17	0
Glucose	100	0.28	0
Sucrose	102	0.24	0
Starch, soluble	100	0.16	0
Methyl acetate	100	0.09	0
Ethyl acetate	108	0.09	0
n-Butyl acetate	101	0.12	0
Benzene	102	0.69	0
Benzoic acid	100	0.40	0
Benzenesulphonic acid	88	0.14	0
Dodecylbenzesulphonic ac.	100	0.40	0
Nitrobenzene	104	0.04	0
Pyridine	50	1.70	0
Acetone	100	0.05	0.01
Toluene	60	1.50	0.10
Aniline	19	4.0	2.40
Phenol	10	4.90	4.00
Salicylic acid	20	3.20	1.50
Ammonia	25	2.10	3.90
Urea	20	4.00	5.20
Thiourea	15	6.50	8.10
DL-Alanine	48	2.20	3.10
L-Glutamic acid	49	1.50	1.80
L-Cysteine	28	4.10	4.90
DL-Methionine	25	2.40	2.60
Cyanide	20	2.70	2.70
Thiocyanate	10	5.00	5.00
Malonic acid	103	0.91	0.91
Acetic acid	100	0	0
Succinic acid	100	0	0

*gram of chlorine per gram of dissolved compound.
Source: [10]

of organics in the wastewater, chlorination often becomes undesirable due to the formation of organohalides, especially trihalomethanes. Under such conditions, other oxidizing agents such as ozone will be needed in order to prevent the formation of organohalides. Ozonation of cyanides and organics is discussed later in this chapter.

Reactions of Chlorine with Organic Compounds in Water

Chlorine is mainly used to disinfect drinking water. It has the ability to substitute for hydrogen, which leads to the formation of carbon-bound chlorine compounds. When certain organic compounds are present in the drinking water, chlorine will react with them to form organohalides, the presence of some of which may be highly undesirable. [13]. For instance, when chlorine reacts with phenols in water, chlorophenols are produced. This class of compounds has been identified as one of the principal sources of bad taste in chlorinated drinking water. It has also been shown recently that chlorination of water containing organics produces trihalomethanes [1]. Their presence in drinking water may pose a health risk, although there is some controversy over the maximum allowable levels permittable for human consumption. While not widely practiced at this time, it is likely that chlorination of drinking water will eventually be replaced by other oxidation processes, such as ozonation or chlorine dioxide oxidation.

Despite the drawbacks associated with drinking water chlorination, this treatment method is very effective in reducing the levels of many organic compounds like acetone, aromatic compounds having o,p-orienting substitutent groups, and amino acids. In addition, ultraviolet radiation-assisted chlorination can drastically increase the conversion of the organics [10]. Compounds like saturated fatty acids, hydroxy acids, unsaturated carboxylic acids, alcohols, carbohydrates and aldehydes are usually inert to oxidation by chlorine, but show substantial decomposition when irradiated with ultraviolet light during chlorination. For other organics, destruction is assisted by ultraviolet radiation. Table 6.2 lists the chlorine consumption associated with the destruction of some organic compounds both with and without the assistance of ultraviolet radiation [10]. For a wide range of compounds, no chlorine is consumed in the absence of ultraviolet light, implying that no oxidation occurs. In most cases where there is chlorine consumption both with and without ultraviolet radiation, ultraviolet irradiation tends to decrease the chlorine demand.

The addition of some metal ions, such as iron (III), copper (II), cerium (IV), mercury (II) and lead (II), can act as catalysts in ultraviolet - assisted chlorination. These metal ions absorb light near 2537 A, which increases the chlorine consumption by as much as 200%, as seen in Table 6.3.

Chlorine has been shown to effectively oxidize organic compounds in wastewater at concentrations less than 1000 ppm and at ambient temperature and pressure. It may not be very effective at high concentrations of organics, and in such cases other alternative technologies (e.g., incineration or wet-air oxidation) might be better choices. The operating conditions for chlorination of organics are very similar to those for cyanide chlorination, that is, the optimal pH is about 10.6, the optimal retention time is sixty minutes, and the chlorine dosage in the wastestream should range from 10 to 20 mg/l [11]. The concentrations of the organic compounds and chlorine in a wastestream can be measured by gas chromatography. From the influent and effluent concentrations of the organics, the removal efficiency can be determined.

6.2.1.b. Capital and Annual Operating Costs of Chlorination

The investment cost for a chlorination system such as the one shown in Figure 6.1 depends primarily on the volumetric flow rate of the wastestream. Table 6.4 shows the installation and equipment costs (with 25% contingency) for 10 and 30 gal/min systems [8]. The prices reflect 1987 costs using published equipment price indices [12]. The volumes of the reactors are fixed such that the optimal one-hour residence time is achieved in each stage. Table 6.5 and Figure 6.2 show the total installation cost for a range of flow

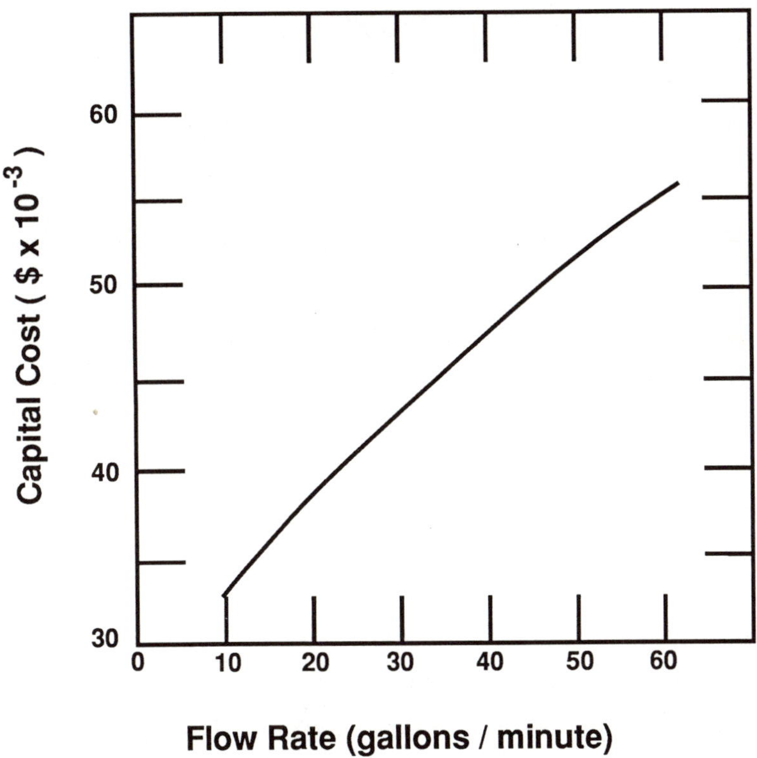

Figure 6.2. Total Installation Costs for Range of Flow Rates

Table 6.3. Effect of Mercury (II) and Lead (II) on Chlorine Consumption of Several Compounds. The Effect Is Calculated by Dividing the Consumption of Chlorine in the Presence of the Catalyst by the Consumption without the Catalyst.

	Effect	
Compound	Hg (II) 10^{-4} M	Pb (II) 10^{-4} M
Oxalic acid	1.5	1.7
Fumaric acid	1.6	1.7
Formaldehyde	2.2	2.0
Methyl alcohol	2.0	1.8
Glycerol	1.5	2.1
Glucose	2.1	1.7
Benzene	1.0	1.1
Salicylic acid	1.1	1.1
DL-Alanine	1.0	1.1

Source: [10]

Table 6.4. Capital Cost for a Two-Stage Chlorination Unit. Installed Cost Includes Field Labor, Equipment Installation, and Shipping.

	Installed Cost	
Item	10 gal/min	33 gal/min
Treatment tank	6400	10600
Reagent storage tanks (2)	2100	3200
Agitators (4)	2100	3200
Pumps (5)	6400	8000
pH controllers/probes (2)	2100	2100
ORP controllers/probes (2)	2500	2500
Piping and valves	2500	3800
Electrical	3200	3200
Total	27300	36600
Contingency (25%)	6800	9100
Total installed cost	34100	45700

Source: [8] with quotes obtained from vendors

Table 6.5. Installed Cost versus Flow Rate for a Two-Stage Chlorination Unit

Q (gal/min)	Capital Cost (in thousands)
10	34.1
20	40.1
30	44.8
40	48.5
50	51.8

Source: [8].

rates. By representing the data in Table 6.5 in terms of the graph in Figure 6.2 it is possible to readily estimate the capital costs of a variety of sizes of installations. Furthermore, it is possible to functionally fit the data in Table 6.5 and obtain a formula which could be used in computerized estimates of the cost of implementing chlorination over a wide range of wastestream sizes.

The total annual operating cost is the sum of capital amortization, fixed and variable operating costs. Assuming an interest rate of 10% and 10 years of annual installments, the cost of annual capital amortization amounts to $5550 for a 10 gal/min unit, and $7440 for a 33 gal/min unit. The fixed operating costs are primarily for electricity, operating labor, maintenance items, and maintenance labor. It is estimated that the fixed operating cost for treatment of cyanide wastes ranges from $2.50 to $7 per pound of cyanide [5], with an average cost of $4.75 per pound of cyanide. The main uncertainty in the operating costs is reagent consumption, which is dependent on the mass flow rates of cyanide and other oxidizable components in the wastestream. The price of chlorine is about $0.11 per pound, and since 6.8 lb of Cl_2 is needed for each pound of cyanide, the cost of chlorine would be $0.75 per pound of cyanide if there are no other oxidizable materials. Therefore, the average total cost of operating a chlorination unit would be $4.75 + 0.75 =$5.50 per pound of cyanide treated.

There are a few companies which manufacture and sell custom-designed chlorination units for treatment of cyanide wastewater for metal finishers. The available units are capable of removing cyanides as well as some heavy metals. The volume of wastewater that the units can handle ranges from 100 gallons per day to 1000 gallons per minute. Table 6.6 provides information on units supplied by two commercial chlorination manufacturers [5]. Because of the large volume of cyanide containing water which usually requires treatment at a typical metal finishing operation, most cyanide wastes are treated on-site. The chlorination unit is located at the source of the wastewater, thereby eliminating the cost of transporting the waste off-site for treatment.

6.2.2. Chlorine Dioxide Oxidation

Oxidation by chlorine dioxide has substantial advantages over oxidation by chlorine in the treatment of drinking water. The oxidation power of chlorine dioxide is not impaired over a wide range of pH, and is therefore not very sensitive to fluctuations in pH. The effectiveness of chlorine oxidation is very much dependent on the pH value as seen in Table 6.1, and at typical operating values of pH from 6 to 10, chlorine partially loses its oxidizing power [3]. Chlorine dioxide does not react with ammonia or nitrogenous compounds, whereas these react with chlorine to form chloramines. Finally, chlorine dioxide does not form trihalomethanes during oxidation of organic chemicals [4]. All of these factors make oxidation by chlorine dioxide one of the techniques of choice for removing low levels of organic chemicals from water.

6.2.2.a. Process Description

Reactions of Chlorine Dioxide with Organic Compounds

Chlorine dioxide does not react chemically with water like chlorine (see Eq. 6.1), and is therefore not suitable to use for cyanide waste treatment. Instead, for treating drinking water, it is used to oxidize iron, manganese and organics, and as well as to disinfect. Most of the oxidizing capacity of chlorine dioxide comes from the reduction of ClO_2 to a chlorite ion through:

$$ClO_2 + e^- \longrightarrow ClO_2^- \qquad (6.6)$$

As with chlorine, wastewater treatment by chlorine dioxide is usually done at the site where the waste is produced. The chlorine dioxide is usually produced by reacting sodium chlorite with acids, such as, hydrochloric or sulfuric acids:

$$5NaClO_2 + 4HCl \longrightarrow 4ClO_2 + 5NaCl + 2H_2O \qquad (6.7)$$

$$10NaClO_2 + 5H_2SO_4 \longrightarrow 8ClO_2 + 5Na_2SO_4 + 4H_2O + 2HCl \qquad (6.8)$$

When using the route represented by Eqn. (6.8) extreme caution must be exercised when mixing $NaClO_2$ with acids. The reaction between sulfuric acid and solid $NaClO_2$ is explosive. Therefore, the reaction (6.8) must be carried out exclusively using solutions of $NaClO_2$ in water.

The by-products produced by the oxidation of most organic compounds with chlorine dioxide tend to be less troublesome in downstream uses of the wastewater than when chlorine is used. Phenol oxidation by chlorine dioxide forms quinones and chloroquinones, which in excess chlorine dioxide, decompose to oxalic and maleic acids [14]. Both products are odorless. This makes chlorine dioxide preferable to chlorine which reacts with phenols to form mono-, di- and trichloro-derivatives. These compounds are highly odorous and only slowly decompose with excess chlorine. Chlorine and chlorine dioxide react similarly with unsaturated aliphatics to form dichloro-compounds, chloroketones, chlorohydrins, and epoxides. Chlorine dioxide oxidizes primary and secondary aliphatic alcohols to acids whereas chlorine does not attack alcohols at all. Table 6.7 summarizes typical products formed when chlorine dioxide is reacted with a variety of organic species [3].

Actual chlorine dioxide oxidation processes utilize operating conditions which are similar to those used in chlorination. The average inlet concentration of ClO_2 can range from 0.1 to 1.5 mg/l, depending upon whether the oxidant is used for bacterial disinfection (low) or for organics removal (high). The efficiency of removal by chlorine dioxide depends upon pH, temperature and pressure. The efficiency increases with increasing pH and pressure, but decreases with increasing temperature [3].

A schematic diagram for a chlorine dioxide system is shown in Figure 6.3. The NaOCl solution is reacted with the $NaClO_2$ solution. The pH is adjusted using hydrochloric acid. The reactions are:

$$NaOCl + HCl \longrightarrow NaCl + HOCl \qquad (6.9)$$

$$HOCl + 2NaClO_2 + HCl \longrightarrow 2ClO_2 + 2NaCl + H_2O \qquad (6.10)$$

Table 6.6. Manufacturers of Commercial Chlorination Units (1986 prices)

Manufacturer	Unit Capacity	# of Units in Service	Cost
Advanced Metal Finishing, Inc.	1000 gallons per day to 350 gallons per minute	>300	20 gpm ....$50000 120 gpm $125000
Advanced Chemical Systems, Inc.	5 gallons per hour to 1000 gallons per minute	50	50 gpm $100000

Source: [5]

Table 6.7 Reactions of Various Organics with Chlorine Dioxide

COMPONENT	END PRODUCTS	TIME (MIN)	pH
3,4 benzopyrene	3,4-benzopyrene-1,5-quinone,8-quinone,10-quinone	60.00	–
aliphatic alcohols	aliphatic acids	–	–
benzylic acid	benzoic acid	–	–
butane-2, 3-diol	acetic acid	–	–
cellotetrose	gluconic groups	240.00	3.00
chlorinated phenols	chloroquinones,halo and non-halogenated aliphatics	–	–
cyclohexene	glutaric,adipic & succinic acids	–	–
cystine	cystine bisulfoxide,cysteic acid	–	3.54
diacetyl butane	acetic acid	–	–
diphenylamine hydrochloride	2-chloro-diphenylamine,4-chloro-diphenylamine	–	–
glucose	aldehydic and carboxylic functional groups	–	2.00
methionine	sulfone	–	–
p-nitrophenol	p-benzoquinone	–	–
pectic acid	galacturonic acid,muric and d,l-tartaric acids	–	–
phenol	oxalic acid,CO_2,H_2O	–	–
thiamine (vitamin b-1)	2-methyl-4-amino-5-aminomethylpyrimidine	–	4.70
triethylamine	acetaldehyde,diethylamine	–	6.00
tyrosine	dopaquinone	–	4.50
vanillin	b-formylmuconic acid,monomethyl ester,diacid	–	4.00
vanillyl alcohol	2-chloro-5-methoxy-1,4-benzoquinone	–	–
veratryl alcohol	4,5-dichloroveratrole	–	–

Source: J. Katz, Ozone and Chlorine Dioxide Technology for Disinfection of Drinking Water Noyes Data Corporation, 1980.

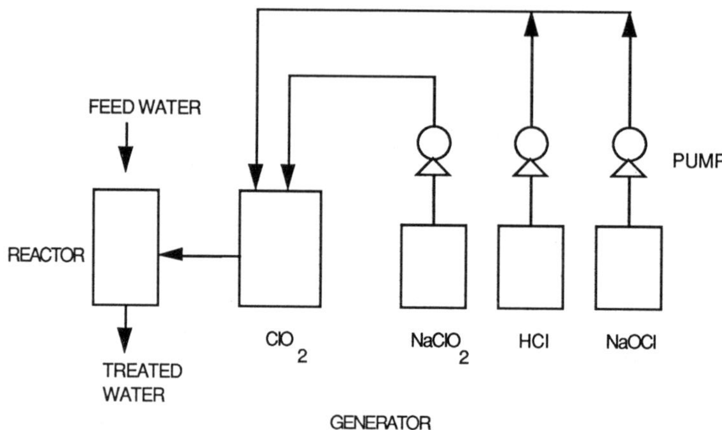

Figure 6.3. Flowsheet for a Typical Chlorine Dioxide Treatment Facility

Typically, a packed bed reactor is used to enhance mass transfer. The flowsheet is arranged so that the acid for both reactions is added simultaneously, with the excess HCl from the first reaction carrying over to react with HOCl in the second reaction. The $NaClO_2$ is added immediately prior to the point where the combined solution enters the packed bed. The performance of the system can be determined by measuring the concentrations of the chlorine dioxide in the entering and exit streams. These concentrations can be found by amperometric titration or spectrophotometry [3]. Monitoring them on - line is essential for establishing an effective control system.

6.2.2.b. Capital and Operating Costs of Chlorine Dioxide Oxidation

The major operating cost of the process diagrammed in Figure 6.3 is the cost of producing chlorine dioxide, which is between $3.00 to $4.37 per pound when hydrochloric acid and sodium chlorite are used as the chemical reagents. The capital cost for the whole system is largely dependent upon: (1) the method of chlorine dioxide production; (2) the production rate; and, (3) the chemical feed systems. The major components are: the reactor, the chemical feed pump(s), the chemical storage tank(s), and the piping/electrical installations. The price of a chlorine dioxide reactor ranges from $1100 to $2000, the price variance being based on reactor capacity, operating conditions (flow rate, residence time, etc.), and valve/piping requirements. Feed pumps that are capable of handling such chemical oxidants generally cost between $670 and $1350. Chemical storage tanks with lids are made from either fiberglass or polyvinylchloride. Steel tanks are seldom used since they are easily corroded. Smaller storage tanks cost from $130 to $840 for capacities of up to 250 gallons. Installation charges can be expected to be approximately one to two times the material cost, which is estimated to be about $7000 to $14000. Table 6.8 summarizes the total capital cost for a chlorine dioxide system. The capital amortization cost, if based on a 10% interest rate and a 10 year payment period, will be $3580 per year.

Table 6.8. Capital Cost for a Ten Million Gallon per Day Chlorine Dioxide System

Item	Cost
Reactor (1)	1550
Pumps (3)	3000
Storage Tanks (3)	1500
Piping and valves	1000
Total material cost	7050
Installed cost (150% of material cost)	10570
Total	17600
Contingency (25% of total cost)	4400
Total installed cost	22000

Source: [3] with quotes (1987) from several U.S. plants.

Since no operating cost data could be found for a chlorine dioxide unit, in all of our estimates, the operating cost of a chlorine dioxide unit is assumed to be the same as the operating cost of a chlorination unit on a per pound of waste treated basis.

6.2.3. Ozonation

Ozone is a powerful oxidizing agent, having oxidizing potential greater than either chlorine or chlorine dioxide. Oxidation by ozone instead of chlorine or chlorine dioxide has been found to eliminate entirely the formation of undesirable end-products such as organohalides. It is effective for disinfection, odor and color removal, and destruction of cyanides and toxic organic compounds in water. In addition, ozonation coupled with ultraviolet radiation destroys some trihalomethanes and their precursors. The unconsumed dissolved ozone after treatment quickly decomposes to oxygen, such that no secondary pollutant is formed which might directly or indirectly (through side reactions) cause environmental health problems.

Ozone is usually generated at the point of use in a flowing air or oxygen stream by an electric discharge process. Mixtures of 1% to 3% of ozone/air and 3% to 5% ozone/oxygen can be produced. These are then mixed with water in a contactor to ensure efficient transfer into the liquid. Figure 6.4 shows a schematic diagram of a typical ozone generating and contacting system for water treatment [1]. Ozone production is energy intensive, with only 10% of the power supplied to the ozone generator producing ozone. The remainder of the power produces light, sound and heat which are undesirable by-products and, hence, represent process inefficiencies. Therefore, the electricity cost for the ozonation process makes up a considerable percentage of the operating cost.

Using oxygen instead of air as the gas feed to the electric discharge unit roughly doubles the ozone production for the same amount of input of electrical energy. However, since only 4% (on the average) of the oxygen feed is converted to ozone, in order to minimize the loss of oxygen, the oxygen coming out from the generator must be recycled to the generator after augmentation with make-up oxygen.

Because ozone is only slightly soluble in water, it is important to design a good contacting system to maximize the mass transfer between the gaseous ozone and the liquid. Some of the factors which affect the mass transfer of ozone are pressure, temperature, bubble size, and method and time of contact [1].

There are no definite limits on the concentrations of the wastes for the ozonation process. Any limitations depend upon operating conditions as well as desired removal efficiencies. However, the removal efficiencies of contaminants are high if the aqueous solutions contain small amounts (up to 1000 mg/l) of contaminants such as organics, cyanides, sulphites, etc. Sludges, slurries and tars are unlikely candidates for ozonation, unless they could be greatly diluted and dispersed in water. Such highly viscous materials pose special mixing problems. Furthermore, they would likely be opaque to ultraviolet radiation limiting the possibility of assisting the oxidation process in this manner (see Section 6.2.3.b).

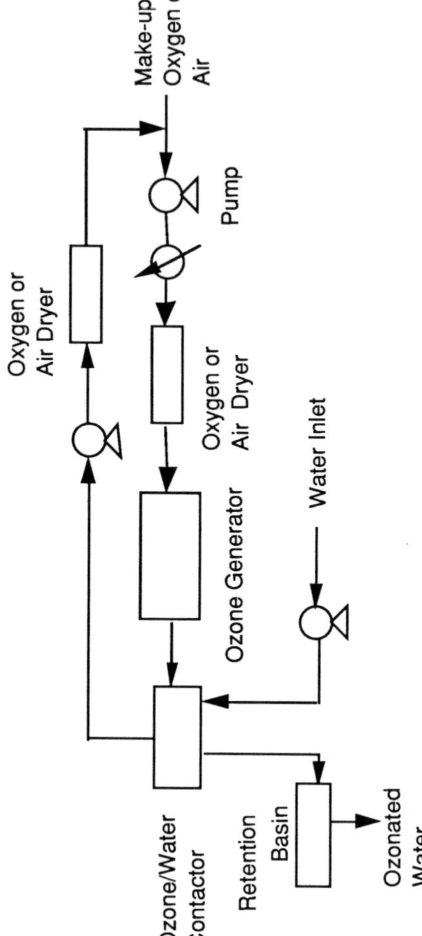

Figure 6.4. Schematic diagram of an ozone generating and contacting system for wastewater treatment.

6.2.3.a Process Description

Removal of Cyanides by Ozonation

Ozone effectively oxidizes cyanides to cyanates. The reaction is normally carried out at ambient temperature and pressure. Ozone oxidation beyond the cyanate level is quite slow and requires another form of oxidation treatment like alkali chlorination in order to convert cyanates to nitrogen and bicarbonates. The reaction of ozone with cyanide is represented by:

$$O_3 + CN^- \longrightarrow CNO^- + O_2 \tag{6.11}$$

Ozone oxidation of cyanide is strongly influenced by pH. The reaction has been found to be first order with respect to the dissolved ozone concentration at all values of pH, but the order for cyanide decreases with increasing pH. The reaction orders with respect to cyanide for pH values of 11, 9.5 and 7 are, respectively, 0.55, 0.83 and 1.06. The optimal pH level for the reaction is seven, with 1.2 moles of ozone being consumed for each mole of cyanide being oxidized [15,16]. Table 6.9 shows an example of removal of cyanides at some pH and ozone concentrations [11]. These data indicate that ozone can usually achieve removal rates of nearly 99% except at the highest concentrations of cyanides. In such cases it is reasonable to expect that a two stage process will be required. These data should be compared with those for cyanide removal by chlorination, Table 6.1. Here, high efficiencies are achieved for a limited range of operating conditions as opposed to the nearly uniformly high efficiency levels indicated in Table 6.9.

The efficiencies of conversion given in Table 6.9 can be determined by measuring the ozone concentration in the influent and effluent streams using the potassium iodide (KI) and the indigo methods [15]. The effect of ultraviolet radiation on ozonation of cyanides is unknown. However, as will be pointed out in the next section, ultraviolet radiation will enhance the oxidation of toxic organic compounds during ozonation. The introduction of copper ions can also catalyze the ozonation of cyanides [5,11]; however, the mechanism for the increase of reaction due to the presence of copper ion is unknown.

Table 6.9. Data on Removal of Cyanides by Ozonation.

pH	Ozone Conc. (mg/l)	Mole ratio (O_3/CN^-)	Cyanide Conc (mg/l) Influent	Effluent	Percent Removal
10.9		0.35-0.5	37.5	0.35	99
9.4		0.35-0.5	75.0	11.50	85
9.5		1.30-1.6	12.9	0.08	99
11.0		1.30-1.6	63.0	0.52	99
7.7		2.30-2.7	13.0	0.02	>99
11.9		2.30-2.7	32.0	0.38	99
9.6		2.30-2.7	34.2	0.60	98
10.8		3.64	29.0	0.08	>99
10.9		2.81	37.5	0.35	99
11.9		2.42	32.0	0.38	99
9.6		2.26	34.2	0.60	98
9.5		2.01	38.4	0.62	98
11.0		1.33	63.0	0.52	99
9.4		0.35	75.0	11.50	85
8.0		11.37	2.1	0.08	96
12.6		6.62	5.3	0.04	99
10.1		5.39	6.5	0.08	99
7.5		1.48	12.9	0.08	99
7.7		2.69	13.0	0.02	>99
9.5		1.05	15.2	0.08	99
7.9	143	2.00	38.4	0.22	99
9.4	143	2.30	34.2	0.34	99
8.9	143	2.30	33.0	0.60	98
	143	2.40	32.8	0.38	99
	143	2.40	32.5	0.60	98
10.1	143	2.40	32.1	0.55	98
11.8	143	2.40	32.0	0.54	98
8.8	143	2.50	31.5	0.52	98
9.1	143	2.60	30.2	0.60	98
	143	2.60	29.5	0.62	98
8.3	64.4	2.70	13.0	0.02	>99
11.3	143	2.70	28.5	0.62	98
9.8	64.4	5.40	6.5	0.08	99
10.8	64.4	6.30	5.6	0.12	98
9.0	64.4	6.60	5.3	0.30	94
11.9	64.4	6.60	5.3	0.04	99
10.0	173	1.33	63.0	0.52	99
8.9	130	1.84	38.0	0.23	99
9.8	195	2.81	37.5	0.35	99
9.8	195	2.91	36.3	0.21	99
8.7	195	3.64	29.0	0.08	>99
7-8	29.7	1.05	15.2	0.08	99
9.5	35.2	1.48	12.9	0.08	99

Source: [11]
Note: Blanks indicate information was not specified.

Reaction of Ozone with Organic Compounds in Water

The main objective of using ozone for oxidation of organic compounds in wastewater is to transform non-biodegradable or refractory components into bio-degradable components. The reaction depends strongly on pH, temperature and rate of ozone decomposition. At low pH values, the ozone molecule, O_3, directly attacks organic substances. Being an electrophilic reagent, ozone is particularly reactive with aromatic compounds like phenols. It is also very reactive with double-bonded compounds, like olefins, to form unstable ozonides which are further decomposed to carboxylic acids and aldehydes :

$$R_1(H)C{=}C(R_2)R_3 + O_3 \rightarrow \text{Ozonide} \rightarrow R_1(H)\overset{+}{C}{-}O{-}O^{-} + O{=}C(R_2)R_3$$

$$R_1(H)\overset{+}{C}{-}O{-}O^{-} \xrightarrow{H_2O} R_1(H)C(O{-}O{-}H)(O{-}H) \;\text{(Hydroxy Hydroperoxide)}$$

$$R_1(H)C(O{-}O{-}H)(O{-}H) \xrightarrow{-H_2O} R_1{-}C({=}O)OH \;\text{(Carboxylic Acid)}$$

$$R_1(H)C(O{-}O{-}H)(O{-}H) \xrightarrow{-H_2O_2} R_1(H)C{=}O \;\text{(Aldehyde)} \qquad (6.12)$$

At high pH values, ozone decomposes to form hydroxyl radical intermediates [1]. Such radicals are more reactive than ozone with certain compounds such as benzene. Table 6.10 gives a list of the products of ozonation of different classes of organic compounds

[17,18]. Table 6.11 gives the removal efficiencies and end-products of specific organic compounds using ozone with or without ultraviolet radiation at ambient temperature and pressure [3,23,24,25]. Listed in Appendix 6.3 are the operating conditions under which the particular removal efficiencies were achieved. Included are data such as influent concentration, residence time and ozone concentration. Such data are particularly useful in the design and scale - up of ozonation processes. It is observed, for example, that the percentage removal by ozonation heavily depends on the residence time of the stream. Only about 10-15% of conversion can be obtained if the residence time is less than five minutes. In order to attain the greatest percentage removal, the residence time should be increased, normally beyond 150 minutes.

The presence of ultraviolet radiation during ozonation greatly enhances the oxidation rate of organic compounds in water. Moreover, ultraviolet radiation during ozonation can remove certain chlorinated compounds and pesticides, such as chlorinated phenols, pentachlorophenol, ethylene dichloride, methylene chloride, polychlorinated biphenyls, kepone, aldrin, dieldrin, endrin and DDT [23] (as indicated by the components with asterisks (*) in Table 6.11), most of which do not react with ozone alone. Table 6.12 shows the effect of ultraviolet radiation on ozonation [1]. Here it is clearly seen that in the case of two relatively simple molecules, ethanol and acetic acid, ozonation without ultraviolet radiation is almost completely ineffective. Ozonation assisted by ultraviolet radiation allows such molecules to be decomposed, although the residence times required are somewhat large.

Besides ultraviolet radiation, both ultrasound and certain metal ions can catalyze the ozonation reaction [20,21]. The catalytic metal ions are nickel, titanium, chromium and iron. The application of ultrasound together with catalyst ions during ozonation gives the

Table 6.10. Types of Organics That Can Be Treated Using Ozonation

Types of Organics	Oxidized Products
Alkanes	alcohols, ketones, aldehydes, peracids and acids
Olefins	carboxylic acids, aldehydes, ketones
Alcohols and ketones	products of alcohol ozonation are aldehydes and ketones, while ketones ozonation yields acids
Aldehydes	peracids, acids and other aldehydes
Acids	carboxylic acids are not reactive with ozone
Benzene and its methylated deriv.	short chain acids, aldehydes
Other benzene derivatives and aromatic comps.	alcohols, aldehydes, ketones, acids, peracids, ozonides and epoxides
Phenols	carbon dioxide, formic acid, glyoxal and oxalic acid
Amines	converted primary and sec. amines to nitro compounds and nitroxides
Pesticides	only aldrin and heptachlor react quantitatively, others are untouched. However, ozonation with ultraviolet radiation can further destroy dieldrin, endrin and DDT.

Source: [17] and [18]

Table 6.11. End Products for Various Organics Treated with Ozone

COMPONENT	END PRODUCTS
* 1,1,1-trichloroethane	--
1,2,3-xylenols	diacetyl,glyoxal,hydrophthalic acid,ketoaldehydes
1,2,4-xylenols	diacetyl,glyoxal,hydroxyphthalic acid,ketoaldehydes
* 1,2-dibromo,3-chloropropane (dbcp)	--
1,4-dichlorobutane	--
1-decene	aldehyde, methylacetal, methyl ester, hydrocarbon
1-propanol	propionaldehyde
2,4-dinitrophenol	--
3-chlorophenol	chloride ion
2,4,5-t	oxalic, glycolic & dichloromaleic acids, chloride ion, CO_2
2,4,6-trichlorophenol	chloride ion
2,4-dichlorophenol	formic acid,oxalic acid, chloride ions
2,4-dinitrophenol	--
2-chlorophenol	chloride ion
2-nitro-p-cresol	nitrate ions
2-propanol	acetone, then acetic and oxalic acids
3,4-benzopyrene	--
3-chlorophenol	chloride ion
3,8-pyrenequinone	1,2,3,4-benzenetetracarboxylic acid, acetic acid
4-aminobenzoic acid	formic & oxalic acids, ammonium nitrate

(continued)

Table 6.11. (continued)

COMPONENT	END PRODUCTS
4-chloro-o-cresol	methylglyoxal,pyruvic,acetic,formic & oxalic acids,CO_2,Cl^- ion
4-chlorophenol	chloride ion
acetic acid	carbon dioxide
acetone	formic and oxalic acids
* aldrin	--
* anisole	--
barium cyanide	N_2, CO_2
benzene	glyoxal, glyoxalic acid, oxalic acid
benzoic acid	other products, CO_2
* bromodichloromethane	--
* bromoform	--
butanol	aldehydes, then acids
cadmium cyanide	N_2, CO_2
carbon disulfide	--
* chlorobenzene	CO_2, oxalic and formic acids
chlorocresols	o-,m-,p-chlorophenols, aliphatic products
chloroform	--
chlorophenol	o-,m-,p-chlorophenols, aliphatic products
cobalt cyanide	N_2, CO_2
copper cyanide	N_2, CO_2
cyanide	N_2, CO_2
cyclohexene	adipic acid, d-hydroxyvaleric acids & its lactone
cysteine	cystine

(continued)

Table 6.11. (continued)

COMPONENT	END PRODUCTS
ddt	--
dieldrin	--
diethyl ether	ethyl acetate, acetate ion
diethylamine	acetaldoxime
diethylbenzene	--
dihydroxyfumaric acid	oxalic acid, mesoxalic acid
dimethyl mercury	--
dinitro-o-cresol	--
* endrin	--
ethanol	aldehydes acetaldehyde, acetic acid, dihydroperoxide
* ethylene dichloride	--
* formic acid	--
gasoline	--
glycerine	--
* glycerol	--
* glycine	--
glyoxal	glyoxalic acid, oxalic acid
gold cyanide	N_2, CO_2
* hexachlorobenzene (hcb)	--
humic acids	formic, acetic and oxalic acids
humic materials	water soluble, ozone-resistant acids
hydrazine	--
hydrogen cyanide	N_2, CO_2
hydrogen sulfide	--

(continued)

Table 6.11. (continued)

COMPONENT	END PRODUCTS
hydroquinone	--
indole	o-aminobenzaldehyde, o-aminobenzoic acid
iron cyanide	N_2, CO_2
isoamyl alcohol	--
* kepone	--
lead cyanide	N_2, CO_2
lindane	--
malathion	degraded products and phosphoric acid
maleic acid	oxalic acid, CO_2, water
malonic acid	hydroxymalonic acid, ketomalonic acid, oxalic, mesoxalic & tartaric acids, CO_2, water
mek	acetate ion
methanol	--
methyl oleate	--
* methylene chloride	--
methylparathion	--
muconic acid	2-carbon, 3-carbon fragments
N, N -diethyl-m-toluamide	formic, acetic & oxalic acids
N, N -diphenylhydrazine hydrochloride	ring- and n-hydroxylated derivatives
naphthalene	salicylic acid
naphthalene-2,7-disulfonic acid	formic, oxalic, mesoxalic acids, sulfate ion, org.carbonyl comp.
naphthols	o-,m-,p-chlorophenols, aliphatic products
nickel cyanide	N_2, CO_2

(continued)

Table 6.11. (continued)

COMPONENT	END PRODUCTS
o,p-xylenols	maleic acid,acetic acid,propionic acid
o-chlorophenol	chloride ion, chlorinated aliphatics
o-cresol	maleic acid,acetic acid,propionic acid
o-toluidine	acetic and oxalic acids
octanol	aldehydes, then acids
oxalic acid	mesoxalic acid, formic acid
p-toluene sulfonic acid	organic peroxides
parathion	2,4-dinitrophenol, picric, sulphuric & phosphoric acids
* pcb	--
* pentachlorophenol (pcp)	--
* perchloroethylene (pce)	--
peroxydiesters	dialcohols, aldehydes
petroleum	--
phenanthrene	2-formylbiphenylcarboxylic,diphenide & diphenic acids
phenol	catechol,p-quinone,cis-muconic,oxalic & fumaric acids, hydroquinone
phosalone	ethers, alcohols
phthalic acid	oxalic & acetic acids, CO_2, water soluble acids
polyhydroxyphenols	o-,m-,p-chlorophenols, aliphatic products
potassium copper cyanide	N_2, CO_2
potassium cyanide	N_2, CO_2
potassium ferricyanide	N_2, CO_2
potassium nickel cyanide	N_2, CO_2
potassium silver cyanide	N_2, CO_2

(continued)

Table 6.11. (continued)

COMPONENT	END PRODUCTS
propionic acid	--
pyrene	acetic & oxalic acids, CO_2, water soluble acids
salicylic acid	phenol, catechol
silver cyanide	N_2, CO_2
skatole	o-aminobenzaldehyde, o-aminobenzoic acid
* sodium acetate	--
sodium cyanide	N_2, CO_2
sodium ferrocyanide cyanide	N_2, CO_2
styrene	benzoic acid, CO_2, H_2O
tartaric acid	dihydroxy-tartaric, glyoxal, oxalic, mesoxalic acids, H_2O_2
tartronic acid	mesoxalic acid
* tetrahydrofuran (thf)	--
thiophenols	o-,m-,p-chlorophenols, aliphatic products
toluene	--
* trichloroethylene (tce)	--
trichloromethylparathion	--
* trinitrotoluene (tnt)	--
* vinyl chloride	--
xylene	--
zinc cyanide	H_2O

The asterisk (*) denotes ultraviolet - assisted treatment.

best overall performance. The order of chemical-oxygen-demand (COD) reduction obtained using one or more catalytic agents is [21]:

sonocatalytic ozonation > sono-ozonation > catalytic ozonation > ozonation
(ozonation + Ni catalyst +ultrasound) (ozonation + ultrasound) (ozonation + Ni catalyst) (6.13)

Table 6.12. Effect of Ultraviolet Radiation on Ozonation

Component	TOC Influent mg/l	TOC Effluent mg/l	UV used	Ozonation Time, min.
Ethanol	41	40	no	240
Ethanol	41	5	yes	240
Acetic acid	38	37	no	90
Acetic acid	38	10	yes	90

Source: [19]

By using sonocatalytic ozonation, removal of 85% of COD can be obtained in two hours, whereas only 70% of COD removal can be achieved by catalytic ozonation, indicating that ultrasound has a substantial influence on the ozonation reaction rate.The mechanism for this is not well - understood, although it might be speculated that high intensity ultrasound puts molecules into a more reactive state which then permits higher degrees of conversion to occur. Sonocation could also diminish any mass transfer limitations, although it is unlikely that such mechanisms are active in the systems considered here. Ozonics Technology in Closter, New Jersey, has patented a process using ultrasound in ozonation.

6.2.3.b. *Capital and Annual Costs of Ozonation/Ultraviolet Ozonation*

The capital cost of an ozonation system includes the cost of ozone generator, air compressors, associated instrumentation, contactors, housing, power supplies, and installation. In addition, the cost of ultraviolet lamps must be considered for ultraviolet - assisted ozonation. The capital cost is very sensitive to the volumetric flow rate of the wastestream, with ozonation being very costly for small capacity generators. The operating cost includes capital amortization, electricity cost, labor and maintenance costs. Tables 6.13 and 6.14 give the capital and annual operating costs of ozonation for small and large wastestreams, respectively [8,22] (all costs are up - dated using the price indices in [12]. The capital amortization is based on 10% annual interest rate and installments for 10 years.

The capital and operating and maintenance costs for small and large capacity ultraviolet-assisted ozonation units are listed in Table 6.15. These data are based upon information received from Ultrox International (personal communication), a company that specializes in ultraviolet assisted ozonation units for wastewater treatment. Comparing the data in Table 6.15 with those in Tables 6.13 and 6.14, which are for ozonation only, it is found that the cost of ultraviolet lamps for ultraviolet assisted ozonation is not very great.

Of the three oxidation processes using chlorine, chlorine dioxide, and ozone, the latter is the most expensive. However, ozonation can eliminate the production of halogenated organic compounds during oxidation, such that no additional treatment is required to remove such products. In some cases, particularly when the waste stream contains either high levels or complex mixtures of organic compounds, this benefit may compensate for the high cost of ozonation.

Table 6.13. Estimated Cost of Ozonation for Small Capacity Generators in Dollars

Type of Cost	Capacity 10 gal/min	30 gal/min
Capital (includes 30% overhead)	109,200	130,000
Capital amortization (10%, 10 yrs)	17,800	21,200
Labor	4,100	6,200
Maintenance (4% of investment)	3,400	4,000
Electricity	550	1,900
Total annual operating cost	25,850	33,300

Source: [8]

Table 6.14. Estimated Cost of Ozonation for Large Capacity Generators in Thousands of Dollars

Type of Cost	Capacity 1 MGD	10 MGD	50 MGD
Ozone generators	236	1933	8697
Compressors and dryers	58	452	1682
Mixers and pumps	68	246	1330
Reactors	176	502	1130
Piping and electrical	146	693	2686
Building and supports	151	427	2159
Overhead (30% of total material cost)	250	1276	5305
Total capital cost	1085	5529	22989
Capital amortization (10%, 10 yrs.)	177	900	3741
Electrical power	65	522	2146
Oxygen	28	110	399
Operation and Maintenance	40	118	301
Total annual operating cost	310	1650	6587

Source: [22]

Table 6.15. Costs of Commercial Ultraviolet/Ozonation Units

Capacity (gal)	Flow Rate (gpm)	Capital Cost (in thousands)	Op. and Maintenance Cost
725	5-50	125-200	$ .2-.25/1000 gal
4800	5-1000	300-1000	$ .2-.25/1000 gal
All prices are in 1987			

Source: [25] with quotes obtained over the phone.

6.3. ASSESSING THE TREATABILITY OF CALIFORNIA WASTES BY CHEMICAL OXIDATION

In Section 6.2, we discussed the applicability of three chemical oxidation processes, using chlorine, chlorine dioxide and ozone, to the treatment of typical wastestreams. This discussion is based upon currently available off - the - shelf technology which, at least according to the vendors, can be readily implemented to achieve predictable levels of waste removal. This information can now be systematically applied to evaluate the potential impact of such technologies on a wide spectrum of hazardous wastes. As discussed in the Introduction, our model spectrum of wastes are those generated in California in 1985 (see Chapter 2) [26]. The 1985 BGR database provides us with a broad sample for establishing the potential for waste reduction by the widespread application of chemical oxidation technologies.

6.3.1. Selection of Appropriate Chemical Process

The first step in the evaluation of the potential impact of chemical oxidation is to examine each wastestream and determine the appropriate chemical oxidation process. Figure 6.5 shows the flow diagram for a computer program which examines the composition and volume of each wastestream in the BGR database to determine which, if any, of the chemical oxidation processes would be appropriate for treating that stream. The program excludes the wastestreams which were reported as being either "solids" or "sludges". In eliminating such wastes, we assumed that their viscosities would render mixing exceptionally difficult and hence there would be severe mass transfer limitations. It is possible that special equipment could be designed to overcome such difficulties, however, for our studies, we have exclusively explored the use of off - the - shelf technologies, as described by vendors. The program also excludes wastestreams which had less than 85% of water. This criterion was applied because it represents the minimum water content of an aqueous liquid stream. At the same time three databases were created which contain the names of components that are treatable by the three chemical oxidation processes being considered. After excluding those wastestreams containing less than 85% water and those which were either slurries or sludges, the name of each component in each wastestream was checked to determine whether it was contained in any of the three databases. The elements of these databases are given in Appendices 6.1and 6.2 and Table 6.11. To the extent available, these Appendices include removal efficiencies, concentrations of the oxidizer, residence times, pH and the source of the information. This checking procedure was repeated until all of the components listed in each wastestream were examined.

If one or more components in the wastestream is contained in one of the Appendices, say Appendix 6.1 (corresponding to chlorination), and the sum of the concentrations falls in the three previously mentioned concentration ranges (0-0.1%, 0.1%-1%, 1%-5%), the chlorination process is deemed applicable with corresponding grades (grade A for high removal efficiency, grade B for medium removal efficiency and grade C for low removal efficiency). Among the three limits, the 0.1% was the most suitable criterion for process selection because it reflects the concentration of contaminants in wastewater that can be treated by any chemical oxidation process while maintaining a reasonably high removal efficiency. If the wastestream selected for chlorination contains a

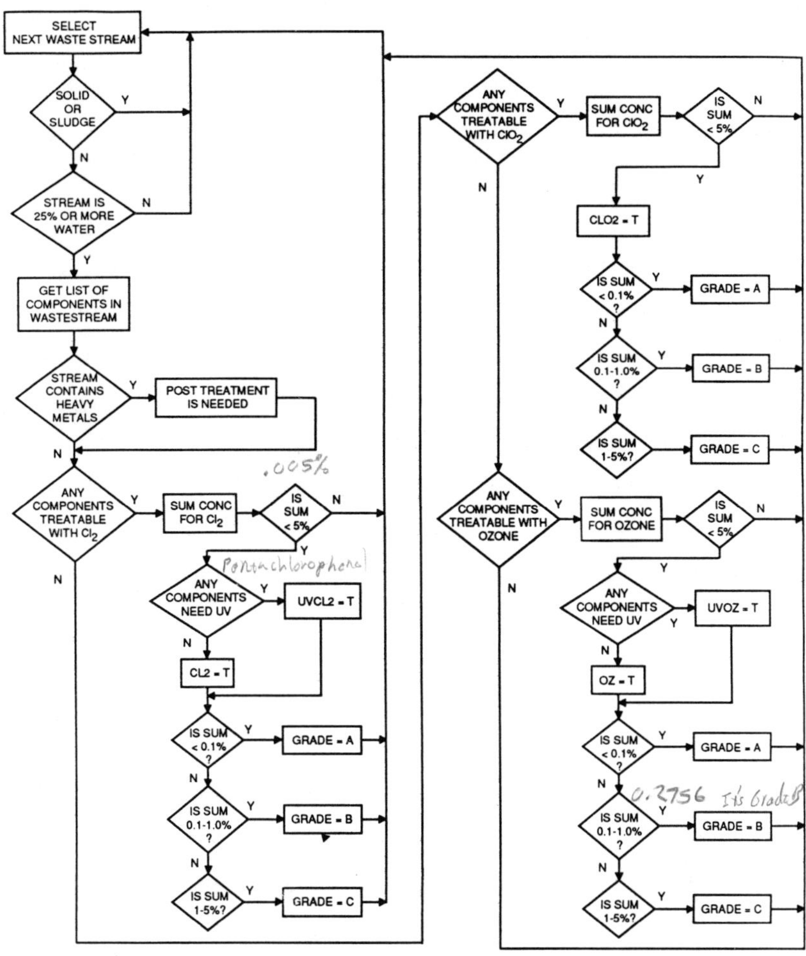

Figure 6.5 Flow Diagram for Selection of Appropriate Chemical Oxidation Processes

component that requires ultraviolet radiation for removal, ultraviolet - assisted chlorination is deemed to be the applicable treatment rather than chlorination alone.

6.3.2. Cost Estimation for Chemical Oxidation Processes

Once the appropriate chemical oxidation process (or processes) was determined, the capital and annual operating costs were calculated from the literature values and vendor information. However, since for each chemical treatment process, cost data were only available for specific process flow rates, certain assumptions were made in order to calculate the capital and annual operating costs for each wastestream which did not have a flow rate that corresponded to any of the available discrete values. The assumptions that were used to compute the capital and annual operating costs are:

1. Since the reported cost data were a function of wastestream flow rate, the annual discharge volume as reported in the Biennial Generator Report Data (see Chapter 2) had to be converted to a volumetric flow rate. We used 250 working days a year and 8 working hours a day as the basis for calculating the wastestream flow rate. Thus, the volumetric flow rate, Q, was calculated using:

$$Q\ (\text{gal/min}) = \frac{\text{Annual Volume(tons/yr)} \times 240\ (\text{gal/ton})}{(250\ \text{day/yr})(8\ \text{hr/day})(60\ \text{min/hr})}$$

$$= (V \times 240)/(250 \times 8 \times 60) = V/500 \qquad (6.14)$$

   where V is the annual volume reported in the BGR database and the density of water (240 gal/ton) is used as the conversion factor from tons to gallons.

2. If the calculated flow rate Q from Eqn. (6.14) is less than or equal to the smallest flow rate for which cost data were found, the capital cost for the smallest flow rate was chosen. For flow rates greater than the smallest one, interpolations or extrapolations were made from the correlation equations of

capital cost versus flow rate obtained by linear or non-linear regressions based upon the available information.

3. The annual operating cost is the sum of capital amortization and the cost of running the process per year (i.e., cost of labor, electricity, chemicals, maintenance, etc.). The annual capital amortization is calculated by assuming a 10-year life time of the equipment and 10% annual interest rate. The formula for the amortization is

$$\text{Cap. Amortization} = \text{Capital Cost} \times [i(i+1)^n]/[(i+1)^n - 1] \qquad (6.15)$$

By substituting

i (interest rate) = 0.1,

n (number of years) =10

into Eqn. (6.15), we find

$$\text{Annual Amortization} = \text{Capital Cost} \times [0.1(1.1)^{10}]/[(1.1)^{10} - 1]$$

$$= 0.1627 \times \text{Capital Cost} \qquad (6.16)$$

The cost of operating the process per year for a given flow rate, Q, can be determined from the correlation of operating cost (excluding capital amortization) versus flow rate, either by extrapolation or interpolation. Therefore, the total operating cost (sum of amortization and cost of operation) can be represented by:

$$\text{Annual Operating Cost} = 0.1627 \times \text{Capital Cost} + \text{Annual Cost of Operation} \qquad (6.17)$$

With the above three assumptions, the capital and annual operating costs for each chemical oxidation process (i.e., chlorination, chlorine dioxide oxidation and ozonation) are computed using the cost estimation flow sheet shown in Figure 6.6.

6.3.2.a. Chlorination/Ultraviolet Chlorination

No cost data for ultraviolet - assisted chlorination could be found; however, since the costs of ozonation with or without ultraviolet irradiation were very close, it is assumed that the costs of ultraviolet chlorination were approximately the same as costs for chlorination without ultraviolet radiation.

Capital Cost

The lowest volumetric flow rate shown in Table 6.4 is 10 gal/min, for which the capital cost is \$34100. Therefore, according to assumption (2), for **Q < 10 gal/min**

$$\text{Capital Cost} = \$34100 \tag{6.18}$$

When Q > 10 gal/min, the correlation between capital cost and flow rate was made by fitting the data in Table 6.4 using a third-degree polynomial. The correlation coefficient is 0.99, indicating a good fit. Thus, for **Q > 10 gal/min,**

$$\text{Capital Cost (\$)} = 26260 + 889.3 \times Q - 11.3 \times Q^2 + 0.075 \times Q^3 \tag{6.19}$$

Annual Operating Cost

The capital amortization is calculated from Eqn. (6.16), using the appropriate capital cost as determined from either Eqn. (6.18) or (6.19). The cost of operating a chlorination process is approximately \$5.50 per pound of components treated. If the annual discharge volume of the wastestream is given in tons, and the concentration of treated components, W_{CL2}, is given in weight percent, then the annual operating, A.O.C., cost will be:

$$\text{A.O.C. (\$/ton)} = 0.1627 \times \text{Capital Cost} + \frac{5.50 \times 2000 \times V \times W_{CL2}}{100}$$

$$= 0.1627 \times \text{Capital Cost} + 110 \times W_{CL2} \times V \tag{6.20}$$

6.3.2.b. Chlorine Dioxide Oxidation

Capital Cost

Only one cost datum could be obtained from the literature [4]. Its reported flow rate is a very large one (10 million gallons per day), and its reported cost was relatively low.

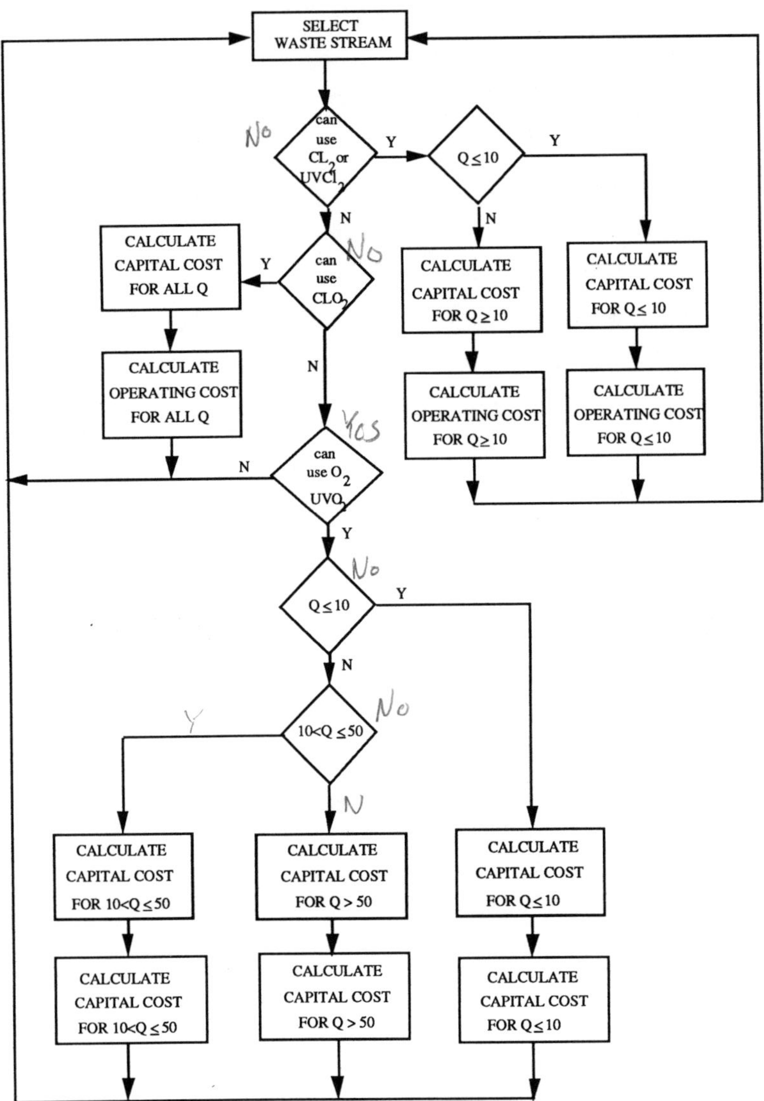

Figure 6.6 Flow Diagram for Cost Estimation for Each Chemical Oxidation Process

Lacking needed data, we have, therefore, assumed a fixed capital cost independent of flow rate. The assumed capital cost for all flow rates is

$$\text{Capital cost} = \$22000 \qquad (6.21)$$

Annual Operating Cost

The capital amortization for the chlorine dioxide oxidation unit amounts to \$3580/yr, which was obtained by substituting the capital cost (\$22000) into Equation 6.15. No reported cost of operating a chlorine dioxide unit could be found; thus it is assumed to be very close to the chlorination case which is about \$5.50 per pound of components being treated. Therefore the annual operating cost equals

$$\text{Annual Operating Cost} = 3580 + 110 \times W_{CLO2} \times V \qquad (6.22)$$

where W_{CLO2} is the concentration of ClO_2-treatable components (in weight percent) in the wastestream.

6.3.2.c. Ozonation

Since the cost data for ozonation with and without ultraviolet are comparable, and the data on ozonation are more complete than in the cases of chlorination and chlorine dioxide oxidation, we used these data in calculating the capital and annual operating costs for both ozonation processes.

Capital Cost

The lowest flow rate shown in Table 6.13 is 10 gal/min. Thus, according to assumption (2), for **Q < 10 gal/min**

$$\text{Capital Cost} = \$109{,}200 \qquad (6.23)$$

For rates falling between 10 and 50 gal/min, the capital cost was estimated from the following linear equation (which is obtained by using the cost data in Table 6.13 at 10 and 30 gal/min flow rates)

$$\text{Capital Cost} = 98800 + 1040 \times Q \qquad (6.24)$$

Almost all of the wastestreams which we wish to analyze have flow rates less than 50 gal/min. Since no cost data were available for flow rates between 30 and 690 gal/min (The cost data in Table 6.14 were for flow rates greater than 690 gal/min.), Eqn. 6.24 was extrapolated to obtain cost estimates for flow rates between 30 and 50 gal/min. This seemed to be a more reasonable assumption than interpolating between flow rates in Tables 6.13 and 6.14 which are more than an order of magnitude apart. The correlation curve for $Q > 50$ gal/min (using Table 6.14) is:

$$\text{Capital Cost} = 5.8\times10^5 + 729.2 \times Q - 2.41\times10^{-3} \times Q^2 \qquad (6.25)$$

Annual Operating Cost

As before, the capital amortization is calculated from Eqn. 6.16, using the appropriate capital cost from Eqn. 6.23, 6.24 or 6.25. As for the cost of operation, if the flow rate is less than or equal to 50 gal/min, a linear relationship between cost and flow rate (using the data in Table 6.13) is assumed. Thus, for $Q < 50$ gal/min, the equation for the annual operating cost takes the form:

$$\text{Annual Operating Cost} = 0.1627 \times \text{Capital Cost} + 6025 + 202.5 \times Q \qquad (6.26)$$

For $Q > 50$ gal/min, the cost data from Table 6.14 are used to determine the correlation equation, which is found to be:

$$\text{Annual Operating Cost} = 0.1627 \times \text{Capital Cost} + 6.12 \times 10^4 + 103.9 \times Q - 6.82\times10^{-4} \times Q^2 \qquad (6.27)$$

6.3.3. Results on Treatability Using Chemical Oxidation

Appropriate process(es) as well as capital and operating costs were determined for wastestreams in the 1985 BGR database. First, the suitable chemical treatment process(es) were determined for a given wastestream, and then the capital and annual operating costs of the process(es) being chosen were calculated. These choices were made using databases containing the list of compounds appropriate for treatment by each of the processes considered (see Appendices). It is important to note that these lists only contain names of components which have been reported to be treatable in the literature. These lists do not necessarily contain all the components that can be treated by such processes. In addition, the concentration limits for the feasibility of the three processes have been set arbitrarily to values of 0.1%, 1% and 5%, which may not be appropriate to all of the components contained in Appendices. Possible interference with the treatment, caused by presence of other components in the wastestream, was neglected. Therefore, the results derived from using these or similar principles should be considered to yield only very rough estimates of the volume of treatable waste.

The results of this study are summarized in Tables 6.16, 6.17 and 6.18 which contain data on specific wastestreams treatable by chlorination, chlorine dioxide oxidation and ozonation, respectively. While these wastestreams are specific to those found in the California BGR survey discussed in Chapter 2, they are indicative of the typical spectrum one might find in any highly industrialized state.

The degree of contaminant removal by each process is indicated by the following three grades: grade A for high efficiency, grade B for medium efficiency and grade C for low efficiency, with respect to their contaminant concentration limits of 0.1%, 1% and 5%. Of the 126 wastestreams treatable by chlorination/ultraviolet chlorination (Table 6.16), 32 were classified as grade A, 44 as grade B and 50 as grade C, indicating that most of the wastestreams in Table 6.16 can only be treated with medium to low removal efficiencies.

Table 6.16. List of Wastestreams Treatable by Chlorination or UV Chlorination

SIC	VOLUME (TONS)	CAPITAL COST ($)	OPERATING COST	CL2	UVCL2	% WATER	GRADE	DESCRIPTION OF GENERATING PROCESS
39*	5.00	34,100	5,570		X	88.00	A	electroless copper plating of printed circuit boards
37	5.00	34,100	5,570	X		89.80	A	spent color chem film for aluminum mixed w/ spent HCl solution
38*	143.60	34,100	7,142	X		89.91	A	waste oil from machines
36	2.00	34,100	5,555	X		90.44	A	thin film head production
36	16.00	34,100	5,566	X		91.13	A	printed circuit board manufacturing
95*	0.83	34,100	5,550		X	91.61	A	damaged through warehouse handling
99	89.00	34,100	6,437	X		93.41	A	substance used for photographic processing
36	12.00	34,100	5,668	X		94.75	A	clarifier sludge
39*	12.00	34,100	5,681		X	95.15	A	deburring of aluminum castings
28*	15,145.00	44,914	24,118		X	95.51	A	manufacture of nitroplasticizers
29*	25.00	34,100	5,798		X	96.24	A	boiler cleaning solution
37*	29.00	34,100	5,709		X	96.35	A	waste machining coolant
38	69.00	34,100	5,701	X		97.97	A	p.c. fabrication: chromate anodize line
25	21.00	34,100	5,571	X		98.48	A	(cu) site cleanup material

(continued)

Table 6.16. (continued)

SIC	VOLUME (TONS)	CAPITAL COST ($)	OPERATING COST	CL2	UVCL2	% WATER	GRADE	DESCRIPTION OF GENERATING PROCESS
38	0.04	34,100	5,548	X		98.97	A	chemical laboratory waste
36	40.00	34,100	5,637	X		98.97	A	aluminum chemical conversion coating
37*	100.83	34,100	5,772	X		98.98	A	fueling operation
07	260.87	34,100	5,838	X		99.17	A	plating shop operation
07	6.00	34,100	5,568	X		99.32	A	pesticide application
34	33.33	34,100	5,622	X		99.69	A	precious metal finishing process containing cyanides
34	20.83	34,100	5,756	X		99.75	A	precious metals finishing process containing cyanides
34	2.08	34,100	5,553	X		99.76	A	precious metal finishing process
34	2.08	34,100	5,550	X		99.77	A	precious metals finishing process containing cyanides
34	6.25	34,100	5,555	X		99.88	A	precious metals finishing process containing cyanides
36	0.06	34,100	5,549	X		99.90	A	circuit board manufacturing
97*	2,841.67	34,100	37,091		X	99.90	A	chemica toilet cleanout
34	14.20	34,100	5,643	X		99.94	A	metal stripping
36	1.15	34,100	5,556	X		99.94	A	semiconductor mfg.
37*	148.49	34,100	6,043		X	99.97	A	machining of rocket engine parts

(continued)

Table 6.16. (continued)

SIC	VOLUME (TONS)	CAPITAL COST ($)	OPERATING COST	CL2	UVCL2	% WATER	GRADE	DESCRIPTION OF GENERATING PROCESS
99*	0.46	34,100	5,549		X	99.97	A	mixed solvent/cleanup material
34	12.50	34,100	5,562	X		99.99	A	process sewer line
49*	3.20	34,100	5,552		X	99.99	A	drain cooling system
29	20.00	34,100	5,992	X		87.33	B	clean up solution from ammonia tank
37*	16.67	34,100	7,398		X	88.00	B	mfg. process
39	71.88	34,100	9,537	X		89.20	B	rinsewater and plating solutions
36*	0.70	34,100	5,626		X	90.00	B	sludge from plating solution
29*	2000.00	34,100	13,762	X		90.40	B	tank cleaning
33	313	34,100	22,572	X		92.65	B	precious metal reclaim
36*	1.83	34,100	5,751		X	93.	B	printed circuit board mfg. process
36	899.17	34,100	71,421	X		93.65	B	spent hydrofluoric acid solution used in semi-conductor mfg.
36	0.48	34,100	5,564	X		93.9	B	electrical and electronic machinery
73	260.	34,100	22,864	X		93.9	B	photographic processing
36	12.	34,100	6,880	X		93.97	B	clean-up of clarifier
28*	12.5	34,100	6,936	X		94.	B	washwater
42*	637	34,100	43,023	X		95.24	B	interior cleaning of bulk liquid trans. equip.
36	19.71	34,100	7,736	X		96.	B	copper etching solution
36*	130.	34,100	12,763	X		96.48	B	etching silicon with hydrofluoric acid

(continued)

Table 6.16. (continued)

SIC	VOLUME (TONS)	CAPITAL COST ($)	OPERATING COST	CL2	UVCL2	% WATER	GRADE	DESCRIPTION OF GENERATING PROCESS
28*	593.	34,100	15,422		X	96.55	B	oilfield chemicals (blending only)
36	1000.	34,100	59,938	X		98.02	B	spent liquid mixture from electro - plating
34	2.08	34,100	5,664	X		98.5	B	spent silver plating solution
37*	363	34,100	21,665		X	98.7	B	spent machine coolant
28*	2.00	34,100	5,770	X		99.00	B	tank bottom sediment
36*	1.6	34,100	6,436	X		85.00	C	electroless tin plating
26*	37.5	34,100	9,711	X		85.	C	chemical and pharmaceutical mfg.
34	37.5	34,100	9,835	X		86.14	C	spent cyanide - zinc plating solution
28*	1006.73	34,100	452,536		X	88.5	C	transfer station
37	20.	34,100	11,098	X		89.00	C	metal finishing activities
28	395.83	34,100	122,860	X		90.93	C	blending and packaging of consumer products for home use
36*	1.5	34,100	5,906	X		91.40	C	positive photoresist process
37	50.	34,100	33,298	X		91.5	C	misc. alkaline waste from labs
99	2.08	34,100	6,702	X		95.	C	printed wiring board production
28	2000	34,100	336,328	X		96.53	C	washing of batching tanks, filling and transfer lines
36*	45.83	34,100	13,179	X		97.70	C	cleaning and stripping of parts

Table 6.17. List of Wastestreams That Can Be Treated by Chlorine Dioxide

SIC	Volume (tons)	Capital Cost	Operating Cost	%Water	Grade	Generating Process
38	143.60	22000	5174	89.91	A	Waste oil from machines
37	148.49	22000	3745	99.97	A	Machining of rocket parts
42	71.47	22000	5801	94.44	B	Interior cleaning of bulk liquid highway transportation equipment
42	637.00	22000	41055	95.24	B	Interior cleaning of bulk liquid highway transportation equipment
28	150.00	22000	11905	98.01	B	Equipment and floor cleaning wash water
37	48.80	22000	4772	99.78	B	Cleaning of waste tank
28	868.48	22000	312064	86.50	C	Waste in manufacturing process
39	4.20	22000	3813	90.00	C	Parts cleaning
36	45.83	22000	8667	97.70	C	Cleaning and stripping of parts

Table 6.18 Typical Wastestreams Treatable by Ozonation

SIC	VOLUME (TONS)	CAPITAL COST ($)	OPERATING COST	OZONE	UVOZONE	% WATER	GRADE	DESCRIPTION OF GENERATING PROCESS
28	34.00	109,200	23,806		X	87.99	A	(cu) site clean-up-removal of underground vessel
37	5.00	109,200	23,794	X		89.80	A	spent color chem film for aluminum mixed w/spent HCl solution
26	20.83	109,200	23,800	X		89.82	A	paper manufacturing and corrugated box plant
38	143.60	109,200	23,850	X		89.91	A	waste oil from machines
37	412.00	109,200	23,959		X	89.99	A	aircraft painting
36	2.00	109,200	23,793	X		90.44	A	thin film head production
36	12.00	109,200	23,797	X		94.75	A	clarifier sludge
37	17.50	109,200	23,799		X	94.85	A	cleaning of waste tank
28	15,145.00	130,302	33,359		X	95.51	A	manufacture of nitroplasticizers
28	2,252.00	109,200	24,704	X		96.21	A	manufacture of 10,10-oxybisphenoxarsine
37	21.00	109,200	23,800		X	97.05	A	manufacture of rocket chambers
25	21.00	109,200	23,800	X		98.48	A	(cu) site cleanup material
38	0.04	109,200	23,792	X		98.97	A	chemical laboratory waste
36	40.00	109,200	23,808	X		98.97	A	aluminum chemical conversion coating
37	100.83	109,200	23,833	X		98.98	A	fueling operation
07	260.87	109,200	23,897	X		99.17	A	plating shop operation
07	104.17	109,200	23,834	X		99.55	A	pesticide application, transportation
34	33.33	109,200	23,805	X		99.69	A	precious metal finishing process containing cyanides
34	20.83	109,200	23,800	X		99.75	A	precious metals finishing process containing cyanides.

(continued)

Table 6.18. (continued)

SIC	VOLUME (TONS)	CAPITAL COST ($)	OPERATING COST	OZONE	UVOZONE	% WATER	GRADE	DESCRIPTION OF GENERATING PROCESS
34	2.08	109,200	23,793	X		99.76	A	precious metal finishing process
34	2.08	109,200	23,793	X		99.77	A	precious metals finishing process containing cyanides
34	6.25	109,200	23,794	X		99.88	A	precious metals finishing process containing cyanides
51	583.33	109,200	24,028		X	99.90	A	contaminated rainwater
36	0.06	109,200	23,792	X		99.90	A	circuit board manufacturing
29	277.00	109,200	23,904		X	99.90	A	contaminated water cleanup
99	4.17	109,200	23,794		X	99.91	A	
37	250.00	109,200	23,893		X	99.93	A	clean-up clarifiers serving water wash paint booths,steam cleaner
34	14.20	109,200	23,798	X		99.94	A	metal stripping
36	1.15	109,200	23,792	X		99.94	A	semiconductor mfg.
28	12.50	109,200	23,797		X	99.95	A	adhesive mixer wash
37	148.49	109,200	23,852		X	99.97	A	machining of rocket engine parts
35	54.38	109,200	23,814		X	99.97	A	34,35 monitoring well sampling and testing
35	60.42	109,200	23,816		X	99.97	A	34,35 monitoring wells pumping, testing
99	0.46	109,200	23,792	X		99.97	A	mixed solvent/cleanup material
34	12.50	109,200	23,797	X		99.99	A	process sewer line
36	11.10	109,200	23,796		X	99.99	A	(cu) spill
37	16.67	109,200	23,799	X		88.00	B	mfg. process
39	71.88	109,200	23,821	X		89.20	B	rinsewater and plating solutions
29	1,000.00	109,200	24,197	X		89.55	B	spent caustic from alkylation operations

(continued)

Table 6.18. (continued)

SIC	VOLUME (TONS)	CAPITAL COST ($)	OPERATING COST	OZONE	UVOZONE	% WATER	GRADE	DESCRIPTION OF GENERATING PROCESS
29	91.00	109,200	23,829	X		90.00	B	mfg of asphalt emulsions product samples, packing leaks
39	26.00	109,200	23,802		X	90.00	B	
28	120.00	109,200	23,840	X		90.78	B	cleanup water from water based primer paint line
33	313.00	109,200	23,919	X		92.65	B	precious metal reclaim
36	0.48	109,200	23,792	X		93.90	B	electrical and electronic machinery
28	12.50	109,200	23,797	X		94.00	B	washwaters from subdivision/packaging line
42	71.47	109,200	23,821	X		94.44	B	interior cleaning of bulk liquid highway transportation equip.
34	500.00	109,200	23,994	X		94.84	B	steam cleaning booth for assemblies
42	637.00	109,200	24,050	X		95.24	B	interior cleaning of bulk liquid highway transportation equip.
33	3,309.00	109,200	25,132	X		96.00	B	oil/water emulsion from press hydraulic system
36	130.00	109,200	23,844	X		96.48	B	etching silicon material with hydrofluoric acid
28	593.00	109,200	24,032	X		96.55	B	oilfield chemicals (blending only)
34	0.25	109,200	23,792	X		97.00	B	metal plating
38	79.00	109,200	23,824	X		97.09	B	3 water curtain spray paint booths
34	15.00	109,200	23,798	X		97.14	B	electroplating
99	5.00	109,200	23,794	X		97.30	B	plating rinses/ spent stripping solution
34	2.08	109,200	23,793	X		97.99	B	metal coating treatment
28	25.00	109,200	23,802	X		98.00	B	solution coating of synthetic substrates
39	12.50	109,200	23,797		X	98.00	B	carbon adsorber condensate

(continued)

Table 6.18. (continued)

SIC	VOLUME (TONS)	CAPITAL COST ($)	OPERATING COST	OZONE	UVOZONE	% WATER	GRADE	DESCRIPTION OF GENERATING PROCESS
28	37.60	109,200	23,807	X		98.01	B	steam regeneration of spent activated carbon
28	150.00	109,200	23,853		X	98.01	B	equipment and floor cleaning wash water
99	0.62	109,200	23,792	X		98.19	B	metal cleaning/stripping
28	14.00	109,200	23,798	X		98.48	B	manufacturing of dispersions
34	2.08	109,200	23,793	X		98.50	B	spent silver plating solution
37	363.00	109,200	23,939	X		98.70	B	spent machining coolant
99	1.04	109,200	23,792		X	99.00	B	rainwater collected in drums previously containing trichloroeth.
34	0.04	109,200	23,792	X		99.00	B	metal plating
99	45.00	109,200	23,810	X		99.00	B	natural seepage
97	41.67	109,200	23,809	X		99.00	B	slop collection tank
36	2.75	109,200	23,793	X		99.00	B	silver plating process for copper bus bar
48	25.00	109,200	23,802	X		99.00	B	pumpings from manholes
48	24.00	109,200	23,802	X		99.00	B	pumpings from manholes
28	29.00	109,200	23,804	X		99.00	B	tank bottom sediment
28	2.00	109,200	23,793	X		99.00	B	tank bottom sediment
28	23.00	109,200	23,801	X		99.50	B	from manufacturing activity of organic pesticide chemicals
36	0.01	109,200	23,792	X		99.50	B	electronic components
13	142.00	109,200	23,849	X		99.50	B	oil production operations
36	0.05	109,200	23,792	X		99.66	B	electronics assembly
37	241.67	109,200	23,890	X		99.70	B	waste coolant from machining of parts
48	11.67	109,200	23,797	X		99.78	B	waste water removed from a vehicle steam cleaning sump
37	48.80	109,200	23,812		X	99.78	B	cleaning of waste tank
28	162.20	109,200	23,858	X		99.80	B	under water clean-up

(continued)

Table 6.18. (continued)

SIC	VOLUME (TONS)	CAPITAL COST ($)	OPERATING COST	OZONE	UVOZONE	% WATER	GRADE	DESCRIPTION OF GENERATING PROCESS
26	37.50	109,200	23,807	X		85.00	C	chemical and pharmaceutical mfg.
36	180.10	109,200	23,865	X		85.86	C	semiconductor manufacturing
34	37.50	109,200	23,807	X		86.14	C	spent cyanide - zinc plating solution
28	868.48	109,200	24,144	X		86.50	C	waste in manufacturing process
34	6.25	109,200	23,794		X	86.50	C	tin stripping
34	4.17	109,200	23,794	X		88.58	C	metal cleaning and plating
37	20.00	109,200	23,800	X		89.00	C	metal finishing activities
25	35.41	109,200	23,806	X		90.00	C	water wash paint booths, spray coating wood tv cabinets
37	53.24	109,200	23,813	X		90.00	C	machine shop
39	4.20	109,200	23,794		X	90.00	C	parts cleaning
36	1.50	109,200	23,792	X		91.40	C	positive photoresist process
13	8000.00	115,440	28,047	X		91.84	C	oilwell drilling
13	4000.00	109,200	25,412	X		91.84	C	oilwell drilling
13	400.00	109,200	23,954	X		91.84	C	oilwell drilling
37	14.20	109,200	23,798	X		93.00	C	test operations
26	37.50	109,200	23,807	X		93.00	C	chemical and pharmaceutical manufacturing
99	0.46	109,200	23,792	X		94.00	C	printed wiring board production
99	1182.00	109,200	24,271	X		94.25	C	waste water generated from the equipment wash rack area
36	0.50	109,200	23,792	X		94.34	C	photographic developing process
36	1280.00	109,200	24,310		X	94.63	C	rainwater collection-chemical vaults

(continued)

Table 6.18. (continued)

SIC	VOLUME (TONS)	CAPITAL COST ($)	OPERATING COST	OZONE	UVOZONE	% WATER	GRADE	DESCRIPTION OF GENERATING PROCESS
28	620.00	109,200	24,043	X		94.68	C	mfg. of speciality chemicals, clarifying of oily waste water
28	104.17	109,200	23,834	X		94.88	C	washing of tanks and loading area
34	0.56	109,200	23,792	X		95.00	C	gold strip
34	3.00	109,200	23,793	X		95.00	C	stripping gold solution
36	0.17	109,200	23,792	X		95.00	C	electroplating electronic components
36	10.00	109,200	23,796	X		95.00	C	metal finishing
37	10.00	109,200	23,796		X	95.24	C	paint stripping
28	14.00	109,200	23,798	X		95.37	C	mfg. of photographic and electronic chemicals
37	52.00	109,200	23,813	X		95.50	C	rinse water from resin impregnation of metal castings
36	2.50	109,200	23,793	X		95.96	C	plating
28	192.50	109,200	23,870	X		96.00	C	tank truck washout compounds
36	2.10	109,200	23,793	X		96.00	C	electronics assembly
28	7066.67	113,499	27,353		X	97.00	C	cleaning solvents
37	6.00	109,200	23,794		X	97.49	C	rinse water from organic paint stripper
38	0.04	109,200	23,792	X		97.50	C	gold stripping
36	10.00	109,200	23,796	X		97.50	C	electroplating of electronic parts
36	45.83	109,200	23,810	X		97.70	C	cleaning and stripping of parts
99	13.33	109,200	23,797	X		97.75	C	silver recovery
39	0.04	109,200	23,792	X		98.00	C	disposal of obsolete stock
39	115.83	109,200	23,839		X	98.45	C	holding tank for etch room wastes
36	4.17	109,200	23,794	X		98.50	C	plating, absorbent pillows

About 40% of the wastes which can be treated by chlorination/ultraviolet chlorination are related to the metal-finishing industry with SIC codes 34 and 36. Since the wastes from such industries often contain cyanides at levels that can be effectively removed by chlorination, this quantitative result is consistent with our qualitative knowledge of waste treatment practices. The remaining wastestreams contain organic compounds or solvents which can be treated with chlorine. However, since chlorine can react with organic compounds to form undesirable halogenated compounds, particularly trihalomethanes, the use of chlorination for treatment of such wastes should be avoided. We have indicated those wastestreams containing organic compounds which may react with chlorine to form undesirable organohalides in Table 6.16 with asterisks (*). Table 6.16 also indicates whether chlorination with or without ultraviolet irradiation is needed for the wastestream, as shown by an "X" in the CL2 column (chlorination without ultraviolet) or UVCL2 column (ultraviolet/chlorination). In all but one case, the capital costs shown in Table 6.16 are the minimum capital costs which we could calculate from our data. This indicates that all the treatable wastestreams had small flow rates (< 10 gal/min.). Similar results on capital costs were obtained for chlorine dioxide oxidation and ozonation (see Tables 6.17 and 6.18).

The wastestreams that can be treated by chlorine dioxide are shown in Table 6.17. Chlorine dioxide oxidation could be used for only 9 out of 8604 wastestreams. The principal reason for this was the smaller number of components which we determined could definitely be treated by this method.

Ozonation was judged to be nearly as widely applicable to the oxidation of wastes as chlorine. Table 6.18 shows that 125 wastestreams were treatable by either ozonation or ultraviolet - assisted ozonation. Of these, 36 wastestreams were graded A, 45 were graded B and 44 were graded C. There were 43 wastestreams derived from metal-finishing industries (SIC codes 33,34 and 36) which contained cyanides or organics. The selection of ozonation with or without ultraviolet irradiation is indicated by the heading OZONE and

UVOZONE in Table 6.18. All except three of the wastestreams in Table 6.18 were below 10 gal/min, such that the minimum capital cost ($109,200) was used.

6.4. REMARKS ON THE USEFULNESS AND APPLICABILITY OF THE TREATMENT ESTIMATION METHODOLOGY

The techniques we have used to estimate the applicability of chemical oxidation to treat a wastestream require three types of information:

1. Physical state of the wastestream — the physical state can be solid, sludge, liquid or gas. In order for chemical oxidation to be applicable, the wastestream has to be either a liquid or gas with at least 85% water content.
2. List of components and their concentrations — this information is used to determine the appropriate chemical treatment process(es). If any of the components in the wastestream are contained in the databases which contain names of compounds which are treatable by the three chemical oxidation processes and the total concentration of the treatable components falls between the three concentration limits (0.1%, 1% and 5%), the corresponding chemical oxidation process is deemed a suitable treatment process for that wastestream with the degree of removal designated by the three grades (grade A for high efficiency, grade B for medium efficiency and grade C for low efficiency).
3. Annual volume of the wastestream — the annual volume data is used to determine the capital and annual operating costs for the selected chemical oxidation process(es). However, since the cost data are expressed as a function of volumetric flow rates, the discharge volume data were converted to volumetric flowrates using Eqn. (6.14) which assumes continuous discharge of the stream throughout the year with 250 operating days per year and 8 hours per day.

In order to calculate the capital and annual operating costs more accurately, it would be necessary to obtain the following information in addition to the data available from the 1985 California BGR survey (see Chapter 2):

1. Actual waste discharge rate and timing information — this is in response to the difficulty of getting the correct volumetric flow rates as mentioned in (3) in the preceding paragraph. If the overall discharge time of the stream is known, we can calculate the average volumetric flow rate from the division of the waste volume by the total discharge time.

2. pH value of the wastestream — since the effectiveness of chemical oxidation is highest at pH's above pH=10 and since it is significantly affected by the drop of pH, it is necessary to know the pH of the wastestream such that the amount of alkali needed to adjust the pH can be determined exactly. This in turn can provide a better estimate of the cost of alkali in the calculation of the operating cost.

6.5 EMERGING TECHNOLOGIES IN CHEMICAL OXIDATION

Besides chlorination, chlorine dioxide oxidation and ozonation, there are some other chemical processes which recently have gained substantial importance in wastewater treatment and will potentially have significant commercial use in the future. They are briefly described in this Section.

Heavy Metal Catalyzed Peroxidation

Hydrogen peroxide decomposes in the presence of several types of catalysts and ultraviolet irradiation into highly reactive intermediates like hydroxyl radicals [1]. The most common catalyst is ferrous ion, which is referred to as the Fenton reagent [27]. The principal use of hydrogen peroxide in waste water treatment is to remove small amounts of phenols, cyanides, and sulfur compounds [11]. In a basic solution of H_2O_2, the peroxide cleaves the benzene ring of phenol to form muconic acid, then adipic acid [28]. For a given phenol waste, the amount of oxidation is increased for higher peroxide concentration, higher temperature and the presence of oxygen [28].

An experimental study of peroxidation of phenols was performed by FMC Corporation [29], using iron salts as the catalysts. The experimental results revealed a 99.94% reduction of phenols (influent phenol conc. of 500 mg/l) in less than 5 minutes in a solution of 1000 mg/l of hydrogen peroxide. Ferrous sulphate appeared to be the most practical catalyst for the system [29].

Ultraviolet Catalyzed Peroxidation

Ultraviolet radiation can catalyze the chemical oxidation of organic contaminants in water using hydrogen peroxide by its effect upon the organic contaminant and its reaction with the H_2O_2. Under ultraviolet radiation, many organic contaminants undergo changes in their chemical structures, thereby becoming more reactive with chemical oxidants such as hydrogen peroxide.

A commercial process utilizing ultraviolet radiation in hydrogen peroxide has been developed by Peroxidation Systems, Inc.[34] to treat wastewaters containing volatile organics (methyl ethyl ketone, methyl isobutyl ketone, toluene and cyclohexane) and insecticides (sulfolane, ethylene dibromide and dibromodichloropropane (DBCP)). New results related to ultraviolet peroxidation of pesticides are expected to be forthcoming from a collaboration between Peroxidation Systems and the Department of Environmental Toxicology at the University of California at Davis.

Hydrogen Peroxide Catalyzed Ozonation

Nakayama and co-workers [30] have used hydrogen peroxide as a catalyst for increasing the effectiveness of ozonation. Their results have shown that hydrogen peroxide/ozone effectively removed various organic compounds such as phenol, ethanol amine, diethanol amine, benzoic acid, propionic acid and acetone [30].

Recently an Advanced Photo-Oxidation process was developed by Water Management Incorporated to treat low levels of cyanides in wastewater [31]. The process generates a gas, known as Photozone Activated Oxygen, that can remove over 90% of the cyanides in 480 minutes at a pH of 11. The Activated Oxygen has an oxidation potential of 2.6 volts, which compares favorably with ozone (2.07 volts) and chlorine (1.36 volts), and thus can provide several beneficial effects in a wide variety of water and wastewater treatments. In addition to cyanides, it can also destroy some other toxic organic compounds such as trichloroethylene, perchloroethylene and ethylene dibromide [32]. The components of Activated Oxygen are: ozone (66.7%); hydroxyl radicals, OH, (14.7%); hydrogen dioxide, HO_2, (6.3%); hydrogen peroxide, H_2O_2, (5.9%); atomic oxygen, O, (4.4%); and other oxidants (2.0%).

Table 6.19 lists the capital and annual operating costs for treatment of cyanide wastes using Photozone [31]. The vendor (Water Management Inc.) claimed that the cost of Photozone system is lower than the cost of alkaline chlorination system, such that the cost advantages of Photozone, when combined with the elimination of hazards associated with chlorination and the well-oxygenated quality of the effluent from the Photozone system, make it the better alternative for treatment of low levels of cyanides.

Potassium Permanganate Oxidation

Potassium permanganate has been used for destruction of organic residues, especially aromatic compounds, in wastewater, by cleaving the ring structure to form a variety of aliphatic acids [28]. Spicher and Skrinde [33] found that 90% of phenol removal was achieved in less than an hour in an alkaline solution of 7:1 $KMnO_4$-to-phenol ppm ratio. In addition to phenols, aldehydes, amines, aromatic alcohols and keto acids were also readily oxidized. No commercial facilities have yet been established utilizing potassium permanganate as a chemical oxidant for wastewater treatment.

Table 6.19. Capital and Annual Operating Costs for Photozone Oxidation of Cyanides*

Type of Cost	Cost
CAPITAL	$ 538,000
Capital Amortization (10%, 10 yr.)	$ 87,600
Oper. and Maintenance (O & M)	$ 78,200
TOTAL ANNUAL OPERATING COST	$165,800
COST PER POUND OF CYANIDE	$ 8.05

*Treatment of 21600 lbs of Cyanides per year.

Source: [31]

6.6. REFERENCES

1. William H. Glaze, in: Control of Organic Substances in Water and Wastewater, (Bernard B. Berger, ed.), p. 148, Noyes Data Corporation, New Jersey, (1987)
2. H. F. Oehlschlaeger, in: Ozone/Chlorine Dioxide Oxidation Products of Organic Materials, (R.G. Rice, ed. and J.A. Cotruvo), p. 20, Int. Ozone Institute, Ohio, (1978)
3. J. Katz, in: Ozone and Chlorine Dioxide Technology for Disinfection of Drinking Water, New Jersey: Noyes Data Corporation, (1980)
4. G.C. White, in: Handbook of Chlorination, New York: Van Nostrand Reinhold Co.(1972)
5. J. Potter, ed., Alternative Technologies for Recycling and Treatment of Hazardous Wastes, California Department of Health Services, July (1986)
6. C.E. Janson, R.E. Kenson and L.H. Tucket, Treatment of heavy metals in wastewaters. Environmental Progress 1: 212 (1982)
7. H. Silman, Treatment of rinse water from electro-chemical process. Metal Finishing 69: 62 (1971)
8. G.C. Cushnie Jr., in: Electroplating Wastewater Pollution Control Technology, Noyes Data Corporation, New Jersey, (1985)
9. Marshal Sittig, in: Pollution Detection and Monitoring Handbook, Noyes Data Corporation, New Jersey, (1974)
10 T. Kobayashi and T. Okuda, A continuous method of monitoring water quality by chlorine consumption under ultraviolet radiation. Water Res. 6:197 (1972)
11. D.J. De Renzo, in: Pollution Control Technology for Industrial Wastewater, Noyes Publications, New Jersey, (1981)
12. Economic indicators. Chemical Engineering 94: 15 (1987)
13. J.C. Morris, Formation of halogenated organics by chlorination of water supplies. EPA-600/1-75-002, U.S. Environmental Protection Agency, Ohio p 15-18 (1978)

14. A.A. Stevens, D.R. Seeger and C.J. Slocum, in: Ozone/Chlorine Dioxide Oxidation Products of Organic Materials, (R.G. Rice, ed. and J.A. Cotruvo), p. 383, Int. Ozone Institute, Ohio, (1978)

15. M.D. Gurol and W.M. Bremen, Kinetics and mechanism of ozonation of free cyanide species in water. Enviro. Sci. Technol.19: 804 (1985)

16. M. Teramoto, Y. Sugimoto, Y. Fukui and H. Teranishi, Overall rate of ozone oxidation of cyanide in bubble column. J. Chem. Eng. Japan, 14: 111 (1981)

17. H.L. Falk and J.E. Moyer, in: Ozone/Chlorine Dioxide Oxidation Products of Organic Materials, (R.G. Rice ed. and J.A. Cotruvo), Int. Ozone Institute, (1978)

18. P. S. Bailey, in: Ozone in Water and Wastewater Treatment, (F. L. Evans III, ed.), Science Publishers Inc., Michigan (1972)

19. J.D. Zeff, UV-OX process for the effective removal of organics in wastewaters. AIChE Symp. Series, 73, no. 167: 206 (1976)

20. C.G. Hewes and R.R. Davison, Renovation of waste water by ozonation. AIChE Symp. Series, 69, no. 129: 71 (1972)

21. J.W. Chen, Catalytic oxidation in advanced waste treatment. AIChE Symp.Series, 69, no. 129: 61 (1972)

22. C.S. Wynn, B.S. Kirk and R. McNabney, Pilot plant for tertiary treatment of wastewater with ozone. AIChE Symp. Series, 69, no. 129: 42 (1972)

23. Briefing: Technologies applicable to hazardous waste. Report prepared for U.S. Environmental Protection Agency by Metcalf and Eddy, Inc., 1987.

24. M. Amdurer et al., Systems to accelerate in situ stabilization of waste deposits.EPA-540-2-86-002, Envirosphere Co., New York, (1986)

25. Information on UV ozonation supplied by Ultrox Intl., Santa Ana, CA 92704.

26. R. L. Bell, A. P. Jackman and R. L. Powell, a report submitted to the California Dept.of Health Services by Dept. of Chemical Engr., Univ. of California, Davis (1988)

27. C. Walling and R. A. Johnson, J. Am. Chem. Soc., 97 363 (1975)

28. E. Ellsworth Hackman III, in: Toxic Organic Chemicals, Destruction and Waste Treatment, Noyes Data Corporation, New Jersey (1978)

29. Phenols in refinery waste water can be oxidized with hydrogen peroxide. Oil and Gas Journal, (1975)

30. S. Nakayama, K. Esaki, K. Nanba, Y. Taniguchi and N. Tobata. Ozone: Science and Engineering, vol. 1, no. 2, p. 119-131 (1979)

31. M. F. Herlacher and F. R. McGregor, Photozone destruction of cyanide waste at Tinker AFB. Soc. of Automotive Engineers Technical Paper Series, no. 870746, (1987)

32. Technical Bulletin on Advanced Photo-Oxidation, published by Water Management Incorporated Colorado (1988)

33. R. G. Spicher and R. T. Skrinde, Effects of potassium permanganate on pure organic compounds. J. American Water Works Asso., vol. 57, no. 4, p. 472, (1965)

34. Technical Bulletin on UV peroxidation from Peroxidation Systems, Inc., Arizona 85711.

APPENDICES
6.1 - 6.3

Appendix 6.1 Chlorination/ UV Chlorination

COMPONENT	END PRODUCTS	INFLUENT CONC. (MG/L)	EFFLUENT CONC. (MG/L)	% REMOVAL	Cl2 CONC. (MG/L)	Cl2 CON-SUMED (G/G)	TIME (MIN)	PH	REF.
* acetaldehyde	--	98.00	--	--	--	--	--	--	10
acetic acid	--	100.00	--	--	--	--	--	--	10
acetone	--	100.00	--	--	--	0.01	--	--	10
* adipic acid	--	102.00	--	--	--	--	--	--	10
ammonia	--	25.00	--	--	--	3.90	--	--	10
aniline	--	19.00	--	--	--	2.40	--	--	10
barium cyanide	N_2, CO_2	--	--	--	--	--	--	--	--
* benzenesulphonic acid	--	88.00	--	--	--	0.14	--	--	10
* benzoic acid	--	100.00	--	--	--	0.40	--	--	10
cadium cyanide	N_2, CO_2	--	--	--	--	--	--	--	--
cobalt cyanide	N_2, CO_2	--	--	--	--	--	--	--	11, 5
copper cyanide	N_2, CO_2	--	--	--	--	--	--	--	--
cyanide	N_2, CO_2	700.00	--	100.00	--	--	--	--	8, 11
cyanide	N_2, CO_2	32.50	--	100.00	--	--	--	--	8, 11
cyanide	N_2, CO_2	5.10	0.10	98.00	--	--	--	--	8, 11

(continued)

Appendix 6.1 (continued)

COMPONENT	END PRODUCTS	INFLUENT CONC. (MG/L)	EFFLUENT CONC. (MG/L)	% REMOVAL	Cl2 CONC. (MG/L)	Cl2 CON-SUMED (G/G)	TIME (MIN)	PH	REF.
cyanide	--	20.00	--	--	--	2.70	--	--	10
* cyclohexanone	--	102.00	--	--	--	--	--	--	10
* d-tartaric acid	--	100.00	--	--	--	--	--	--	10
dl-alanine	--	48.00	--	--	--	3.10	--	--	10
* dl-lactic acid	--	130.00	--	--	--	--	--	--	10
dl-methionine	--	25.00	--	--	--	2.60	--	--	10
* dodecylbenzenesulphonic acid	--	100.00	--	--	--	0.40	--	--	10
* ethyl acetate	--	108.00	--	--	--	0.09	--	--	10
* ethyl alcohol	--	100.00	--	--	--	--	--	--	10
ethylene	ethylene epoxide	--	--	--	--	--	--	--	3
* ethylene glycol	--	100.00	--	--	--	--	--	--	10
* formaldehyde	--	100.00	--	--	--	--	--	--	10
* formic acid	--	100.00	--	--	--	--	--	--	10
* fumaric acid	--	100.00	--	--	--	--	--	--	10
* glucose	--	100.00	--	--	--	--	--	--	10
* glutaric acid	--	132.00	--	--	--	--	--	--	10
* glycerol	--	100.00	--	--	--	--	--	--	10
* glycolic acid	--	95.00	--	--	--	--	--	--	10

(continued)

Appendix 6.1 (continued)

COMPONENT	END PRODUCTS	INFLUENT CONC. (MG/L)	EFFLUENT CONC. (MG/L)	% REMOVAL	Cl2 CONC. (MG/L)	Cl2 CON-SUMED (G/G)	TIME (MIN)	PH	REF.
gold cyanide	N_2, CO_2	--	--	--	--	--	--	--	11, 5
hydrogen cyanide	N_2, CO_2	6.80	--	100.00	--	--	--	--	11
iron cyanide	N_2, CO_2	0.19	0.08	58.00	20.00	--	30.00	8.80	11
iron cyanide	N_2, CO_2	0.19	0.05	74.00	20.00	--	60.00	8.80	11
iron cyanide	N_2, CO_2	0.19	0.07	63.00	20.00	--	90.00	8.80	11
iron cyanide	N_2, CO_2	0.19	0.04	79.00	10.00	--	30.00	10.60	11
iron cyanide	N_2, CO_2	0.19	0.03	84.00	10.00	--	60.00	10.60	11
iron cyanide	N_2, CO_2	0.19	0.04	79.00	10.00	--	90.00	10.60	11
iron cyanide	N_2, CO_2	0.19	0.03	84.00	20.00	--	30.00	10.60	11
iron cyanide	N_2, CO_2	0.19	0.02	89.00	20.00	--	60.00	10.60	11
iron cyanide	N_2, CO_2	0.19	0.02	89.00	20.00	--	90.00	10.60	11
iron cyanide	N_2, CO_2	0.19	0.03	84.00	10.00	--	30.00	11.00	11
iron cyanide	N_2, CO_2	0.19	0.03	84.00	10.00	--	60.00	11.00	11
iron cyanide	N_2, CO_2	0.19	0.03	84.00	10.00	--	90.00	11.00	11
iron cyanide	N_2, CO_2	0.19	0.01	95.00	20.00	--	30.00	11.00	11
iron cyanide	N_2, CO_2	0.19	0.02	89.00	20.00	--	60.00	11.00	11

(continued)

Appendix 6.1 (continued)

COMPONENT	END PRODUCTS	INFLUENT CONC. (MG/L)	EFFLUENT CONC. (MG/L)	% REMOVAL	Cl2 CONC. (MG/L)	Cl2 CON-SUMED (G/G)	TIME (MIN)	PH	REF.
iron cyanide	N_2, CO_2	0.19	0.02	89.00	20.00	--	90.00	11.00	11
* isopropyl alcohol	--	99.00	--	--	--	--	--	--	10
l-cysteine	--	28.00	--	--	--	4.90	--	--	10
l-glutamic acid	--	49.00	--	--	--	1.80	--	--	10
lead cyanide	N_2, CO_2	68.30	0.10	99.00	--	--	--	--	8, 11
maleic acid	dichloromaleic acid	--	--	--	--	--	--	--	3
malonic acid	--	103.00	--	--	--	0.91	--	--	10
* mek	--	100.00	--	--	--	--	--	--	10
* methyl acetate	--	100.00	--	--	--	0.09	--	--	10
* methyl alcohol	--	100.00	--	--	--	--	--	--	10
* n-acetyl-l-glutamic acid	--	96.00	--	--	--	--	--	--	10
* n-butyl acetate	--	101.00	--	--	--	0.12	--	--	10
* n-butyl alcohol	--	100.00	--	--	--	--	--	--	10
* n-butyric acid	--	130.00	--	--	--	--	--	--	10
nickel cyanide	N_2, CO_2	--	--	--	--	--	--	--	11, 5
* nitrobenzene	--	104.00	--	--	--	0.04	--	--	10
* oxalic acid	--	100.00	--	--	--	--	--	--	10

(continued)

Appendix 6.1 (continued)

COMPONENT	END PRODUCTS	INFLUENT CONC. (MG/L)	EFFLUENT CONC. (MG/L)	% REMOVAL	Cl2 CONC. (MG/L)	Cl2 CON-SUMED (G/G)	TIME (MIN)	PH	REF.
phenol	--	10.00	--	--	--	4.00	--	--	10
phenol	o-chloro-phenol	--	--	--	--	--	--	--	--
potassium copper cyanide	N_2, CO_2	--	--	--	--	--	--	--	11, 5
potassium cyanide	N_2, CO_2	--	--	--	--	--	--	--	--
potassium ferricyanide	N_2, CO_2	--	--	--	--	--	--	--	11, 5
potassium ferrocyanide	N_2, CO_2	--	--	--	--	--	--	--	11, 5
potassium nickel cyanide	N_2, CO_2	--	--	--	--	--	--	--	11, 5
potassium silver cyanide	N_2, CO_2	--	--	--	--	--	--	--	11, 5
* propionic acid	--	89.00	--	--	--	--	--	--	10
* pyridine	--	50.00	--	--	--	1.70	--	--	10
salicylic acid	--	20.00	--	--	--	1.50	--	--	10
silver cyanide	N_2, CO_2	--	--	--	--	--	--	--	11, 5
sodium cyanide	N_2, CO_2	--	--	--	--	--	--	--	--
sodium ferrocyanide	N_2, CO_2	--	--	--	--	--	--	--	11, 5
* sorbitol	--	100.00	--	--	--	--	--	--	10
* starch, soluble	--	100.00	--	--	--	0.16	--	--	10
succinic acid	--	100.00	--	--	--	--	--	--	10

(continued)

Appendix 6.1 (continued)

COMPONENT	END PRODUCTS	INFLUENT CONC. (MG/L)	EFFLUENT CONC. (MG/L)	% REMOVAL	Cl2 CONC. (MG/L)	Cl2 CON-SUMED (G/G)	TIME (MIN)	PH	REF.
* sucrose	--	102.00	--	--	--	--	--	--	10
thiocyanate	--	10.00	--	--	--	5.00	--	--	10
thiourea	--	15.00	--	--	--	8.10	--	--	10
toluene	--	60.00	--	--	--	0.10	--	--	10
unsaturated aliphatics	dichloride	--	--	--	--	--	--	--	3
urea	--	20.00	--	--	--	5.20	--	--	10
zinc cyanide	N_2, CO_2	68.30	0.10	99.00	--	--	--	--	8, 11

Appendix 6.2 Chlorine Dioxide Oxidation

COMPONENT	END PRODUCTS	TIME (MIN)	PH	REF.
3,4 benzopyrene	3,4 - benzopyrene - 1,5 - quinone, 8 - quinone, 10 - quinone	60.	--	3
aliphatic alcohols	aliphatic acids	--	--	3
benzylic acid	benzoic acid	--	--	3
butane -2, 3 - diol	acetic acid	--	--	3
cellotetrose	gluconic groups	240.	3.0	3
chlorinated phenols	chloroquinones, halo and non - halogenated aliphatics	--	--	3
cyclohexane	glutaric, adipic & succinic acids	--	--	3
cystine	cystine bisulfoxide, cysteic acid	--	3.54	3
diacetyl butane	acetic acid	--	--	3
diphenylamine hydrochloride	2 - chloro - diphenylamine, 4 - chloro - diphenylamine	--	--	3
glucose	aldehydic and carboxylic functional groups	--	2.	3
methionine	sulfone	--	--	3
p - nitrophenol	p - benzoquinone	--	--	3
pectic acid	galacturonic acid, muric and d, l - tartaric acids	--	--	3
phenol	oxalic acid	--	--	3
thiamine	2 - methyl - 4 - amino - 5 - aminomethylpyrimidine	--	4.7	3
triethylamine	acetaldehyde, diethylamine	--	6.	3
tyrosine	dopaquinone	-	4.5	3
vanillin	b - formylmuconic acid, monomethyl ester, diacid	--	4.0	3
vanillyl alcohol	2 - chloro - 5 methoxy - 1, 4 - benzoquinone	--	--	3
veratryl alcohol	4, 5 - dichloroveratrole	--	--	3

Appendix 6.3 Ozonation

COMPONENT	infl. conc. (mg/l)	effl.. conc. (mg/l)	% removal	ozone conc. (mg/l)	ozone consumed (g/g)	pH	time (min)	ref.
1,1,1 - trichloroethane	0.17	--	94.1	--	--	--	--	25
1,2,3 - xylenols	--	--	--	--	--	--	--	3
1,2,4 - xylenols	--	--	--	--	--	--	--	3
1,2 - dibromo,								
3 - chloropropane	0.5	--	100.	--	--	--	30	25
1,4 - dichlorobutane	--	--	50.	--	--	--	60	3
1 - decene	--	--	--	--	--	--	960	3
1 - propanol	408	145	65	--	--	9.	60.	3
2,4 - dinitrophenol	50.	0.35	99.3	100.	--	--	--	24
2,4 - dinitrophenol	3.	0.05	98.3	17.	--	--	--	24
2,4 - dinitrophenol	3.	--	100.	14.	--	--	--	24
3 - chlorophenol	--	--	100.	--	1.2	--	--	3
2,4,5 - T	0.26	--	--	--	--	--	60.	3
2,4, 6- trichlorophenol	--	--	100.	--	1.2	--	--	3
2,4 - dichlorophenol	--	--	--	--	--	--	--	3
2,4 - dinitrophenol	50.	0.35	100.	--	2.	--	--	3
2 - chlorophenol	--	--	100.	--	1.2	--	--	3
2 - nitro - p - cresol	--	--	--	--	--	--	--	3
2 - propanol	--	--	85.	--	--	--	135.	3
3,4 - benzopyrene	0.9	0.00	100.	--	0.4	--	7.5	3
3 - chlorophenol	--	--	100.	--	1.2	--	--	3
3,8 pyrenequinone	--	--	100.	--	--	--	1980.	3
4 - aminobenzoic acid	--	--	--	--	--	--	80.	3

(continued)

Appendix 6.3 (continued)

COMPONENT	infl. conc. (mg/l)	effl.. conc. (mg/l)	% removal	ozone conc. (mg/l)	ozone consumed (g/g)	pH	time (min)	ref.
4 - chloro - o - cresol	--	--	100.	--	6.3	--	80.	3
4 - chlorophenol	--	--	100	--	1.2	--	--	1
acetic acid	--	--	14.	--	--	7.	120.	3
acetone	100.	30.	70.	--	--	--	--	3
aldrin	--	--	--	--	--	--	--	23
anisole	--	--	--	--	--	--	--	25
benzene	200.	5.	97.5	20.	--	--	--	3
benzene	500.	--	100	--	--	--	--	24
benzoic acid	300.	45.	85.	--	1.6	--	30.	3
bromodichloromethane	0.1	--	100.	--	--	--	--	25
bromoform	0.1	--	100.	--	--	--	--	25
butanol	--	--	--	33.	--	--	--	3
caffeine	660.	490.	25.	--	1.00	--	90.	3
carbon disulfide	100.	--	100.	--	--	--	--	24
chlorobenzene	--	--	--	--	--	--	--	25
chlorocresols	--	--	--	--	--	--	--	25
chloroform	--	--	50.	--	--	--	60.	3
chlorophenol	--	--	--	--	--	--	--	3
cyanide	37.5	0.35	99.	--	0.87	10.9	--	11
cyanide	75.	11.5	85.	--	0.87	9.4	--	11
cyanide	12.9	0.08	99.	--	2.7	9.5	--	11
cyanide	63.	0.52	99.	--	2.68	11.	--	11
cyanide	13.	0.02	99.	--	4.62	7.7	--	11
cyanide	32.	0.38	99.	--	4.62	11.9	--	11

(continued)

Appendix 6.3 (continued)

COMPONENT	infl. conc. (mg/l)	effl.. conc. (mg/l)	% removal	ozone conc. (mg/l)	ozone consumed (g/g)	pH	time (min)	ref.
cyanide	34.2	0.6	98.	--	4.62	9.6	--	11
cyanide	29.	0.08	99.	--	6.72	10.8	--	11
cyanide	37.5	0.35	99.	--	5.19	10.9	--	11
cyanide	32.	0.38	99.	--	4.47	11.9	--	11
cyanide	34.2	0.60	98.	--	4.17	9.6	--	11
cyanide	38.4	0.62	98.	--	3.71	9.5	--	11
cyanide	63.	0.52	99.	--	2.46	11.	--	11
cyanide	75.	11.5	85.	--	0.65	9.4	--	11
cyanide	2.1	0.08	96.	--	21.	8.	--	11
cyanide	5.3	0.04	99.00	--	12.22	12.6	--	11
cyanide	6.5	0.08	99.00	--	9.95	10.1	--	11
cyanide	12.9	0.08	99.00	--	2.73	7.5	--	11
cyanide	13.	0.02	99.00	--	4.97	7.7	--	11
cyanide	15.2	0.08	99.00	--	1.94	9.5	--	11
cyanide	38.4	0.22	99.00	143.	3.69	7.9	--	11
cyanide	34.2	0.34	99.00	143.	4.25	9.4	--	11
cyanide	33.	0.06	98.00	143.	4.25	8.9	--	11
cyanide	32.8	0.38	99.00	143.	4.43	--	--	11
cyanide	32.5	0.60	98.00	143.	4.43	--	--	11
cyanide	32.1	0.55	98.00	143.	4.43	10.1	--	11
cyanide	32.	0.54	98.00	143.	4.43	11.8	--	11
cyanide	31.5	0.52	98.00	143.	4.62	8.8	--	11
cyanide	30.2	0.6	98.00	143.	4.8	9.1	--	11
cyanide	29.5	0.62	98.00	143.	4.8	--	--	11

(continued)

Appendix 6.3 (continued)

COMPONENT	infl. conc. (mg/l)	effl.. conc. (mg/l)	% removal	ozone conc. (mg/l)	ozone consumed (g/g)	pH	time (min)	ref.
cyanide	13.	0.02	99.00	64.4	4.98	8.3	--	11
cyanide	28.5	0.62	98.00	143.	4.98	11.3	--	11
cyanide	6.5	0.08	99.00	64.4	9.97	9.8	--	11
cyanide	5.6	0.12	98.	64.4	11.63	10.8	--	11
cyanide	5.6	0.12	98.	64.4	11.6	10.8	--	11
cyanide	5.3	0.3	94.	64.4	12.2	9.	--	11
cyanide	5.3	0.04	99.00	64.4	12.2	11.9	--	11
cyanide	63.	0.52	99.00	173.	2.46	10.	--	11
cyanide	38.	0.23	99.00	130.	3.4	8.9	--	11
cyanide	37.5	0.35	99.00	195.	5.19	9.8	--	11
cyanide	36.3	0.21	99.00	195.	5.37	9.8	--	11
cyanide	29.	0.08	99.00	195.	6.72	8.7	--	11
cyanide	15.2	0.08	99.00	29.7	1.94	7.5	--	11
cyanide	12.9	0.08	99.00	35.2	2.73	9.5	--	11
cyclohexene	--	--	--	--	--	--	--	3
cysteine	--	--	--	--	--	--	--	3
DDT	0.5	0.25	50.	13.8	--	--	--	24
DDT	--	--	90.	24.	--	--	--	3
dieldrin	--	--	90.	24.	--	--	--	3
diethyl ether	--	--	94.	--	--	9.	120.	3
diethylamine	--	--	--	--	--	--	--	3
diethylbenzene	55.	0.65	98.00	3.	0.05	--	8.00	3
diethylbenzene	112.5	8.5	92.	8.75	--	--	--	24

(continued)

Appendix 6.3 (continued)

COMPONENT	infl. conc. (mg/l)	effl.. conc. (mg/l)	% removal	ozone conc. (mg/l)	ozone consumed (g/g)	pH	time (min)	ref.
dihydroxyfumaric acid	--	--	--	--	0.45	--	--	3
dimethyl mercury	--	--	100.	--	--	--	10.	3
dinitro - o - cresol	10.	--	100.	5.5	--	--	--	24
endrin	--	--	--	--	--	--	--	23
ethanol	1000.	90.	91.	33	--	--	--	3
ethanol	5000	--	--	--	--	--	120	3
ethylene dichloride	--	--	--	--	--	--	--	25
formic acid	--	--	--	--	--	--	--	25
gasoline	50.	1.	98.	1.29	--	--	--	24
gasoline	1000	--	100	--	--	--	--	24
glycerine	1000	--	100	--	--	--	--	24
glycerol	--	--	--	--	--	--	--	25
glycine	--	--	--	--	--	--	--	25
glyoxal	--	--	100	--	--	--	60	3
hexachlorobenzene	0.5	--	100.	--	--	--	60.	25
humic acids	--	--	--	--	--	--	80.	3
humic materials	--	--	--	--	--	--	--	3
hydrazine	100.	--	100.	--	--	--	--	24
hydrogen cyanide	10.	--	100.	--	--	--	--	24
hydrogen sulfide	100.	--	100.	--	--	--	--	24
hydroquinone	100.	--	100.	--	--	--	--	24
indole	--	--	--	--	--	--	--	3
isoamyl alcohol	1000.	80.	92.	--	--	--	--	24
kepone	--	--	--	--	--	--	--	23

(continued)

Appendix 6.3 (continued)

COMPONENT	infl. conc. (mg/l)	effl.. conc. (mg/l)	% removal	ozone conc. (mg/l)	ozone consumed (g/g)	pH	time (min)	ref.
lindane	--	--	90.	24.	--	--	--	3
malathion	10.	2.	80.	--	0.5	--	--	3
malathion	10.	1.	90.	1.	--	--	--	3
malathion	10.	--	100.	2.6	--	--	--	3
maleic acid	116.	0.95	100.	--	1.68	--	50.	3
malonic acid	--	--	--	--	1.85	4.	90.	3
MEK	--	--	--	--	--	7.	120.	3
methanol	2000.	160.	92.	--	--	--	--	24
methyl oleate	--	--	82.	--	--	--	25.	3
methylene chloride	--	--	--	--	--	--	--	25
methylparathion	10.	0.5	95.	4.5	--	--	--	24
methylparathion	10.	0.1	99.	9.5	--	--	--	24
muconic acid	--	--	--	--	--	--	--	3
N,N-diethyl-m-toluamide	--	--	--	--	--	--	--	3
N,N-diphenyl-hydrazine hydrochloride	--	--	--	--	--	7.	--	3
naphthalene	--	--	--	--	--	--	--	3
naphthalene-2,7-disulfonic acid	--	--	--	--	--	--	120.	3
naphthols	--	--	--	--	--	--	--	3
o,p - xylenols	--	--	--	--	--	--	--	3
o - chlorophenol	--	--	--	--	--	--	--	3

(continued)

Appendix 6.3 (continued)

COMPONENT	infl. conc. (mg/l)	effl.. conc. (mg/l)	% removal	ozone conc. (mg/l)	ozone consumed (g/g)	pH	time (min)	ref.
o - cresol	100.	0.00	100.	--	--	--	--	3
o - toluidine	--	--	--	--	--	--	--	3
octanol	--	--	--	33.	--	--	--	3
oxalic acid	--	--	60.	--	0.65	--	--	3
p - toluene sulfonic acid	--	--	100.	--	--	--	120.	3
parathion	--	--	90.	24.	--	--	--	3
PCB	0.05	--	100.	--	--	--	--	25
pentachlorophenol	0.5	--	100.	--	--	--	15.	23
perchloroethylene	0.96	0.01	98.9	--	--	--	--	25
peroxydiesters	--	--	--	--	--	--	--	3
petroleum	10.	0.25	97.5	4.5	--	--	--	24
phenanthrene	--	--	90.	--	--	--	--	3
phenol	100.	0.00	100.	--	--	--	30.	3
phenol	150.	0.0	100.	25.	0.2	--	--	3
phosalone	--	--	--	--	--	--	--	3
phthalic acid	--	--	100.	--	--	--	1440.	3
polyhydroxyphenols	--	--	--	--	--	--	--	3
propionic acid	490.	235.	52.	--	--	9.	120.	3
pyrene	--	--	100.	--	--	--	1440.	3
salicylic acid	150.	--	100.	--	0.38	--	--	3
skatole	--	--	--	--	--	--	--	3
sodium acetate	--	--	--	--	--	--	--	25
styrene	--	--	--	--	--	--	--	3

(continued)

Appendix 6.3 (continued)

COMPONENT	infl. conc. (mg/l)	effl.. conc. (mg/l)	% removal	ozone conc. (mg/l)	ozone consumed (g/g)	pH	time (min)	ref.
tartaric acid	--	--	--	--	--	5.	80.	3
tartronic acid	--	--	--	--	0.4	--	40.	3
tetrahydrofuran	5.	--	100.	--	--	--	--	25
thiophenols	--	--	--	--	--	--	--	3
toluene	500.	--	100.	--	--	--	--	24
trichloroethylene	0.47	0.01	97.9	--	--	--	--	25
trichloromethyl - parathion	0.75	--	100.	4.	--	--	--	24
trinitrotoluene	--	--	--	--	--	--	--	25
vinyl chloride	--	--	--	--	--	--	--	25
xylene	500.	--	100.	--	--	--	--	24